To our children:
Sabine and Darius

Table of Contents

About the Authors

David A. Lovejoy is Professor of Neuroendocrinology in the Department of Cell and Systems Biology at the University of Toronto in Toronto, Canada. Previous to taking his appointment at the University of Toronto he was a Lecturer at the University of Manchester in Manchester, England. He is an author of over seventy scientific publications in the field of reproductive and stress-related physiology and is the author of the book *Neuroendocrinology, An Integrated Approach*.

Dalia Barsyte is currently a senior scientist at the Structural Genomics Consortium at the University of Toronto and is an author of numerous publications in the field of the molecular biology of stress, cancer and environmental toxicology.

Preface

This book is intended to provide an understanding of how mechanisms of reproduction and stress-related physiology interact to allow organisms to cope and survive in an all-too-frequently hostile environment. Although it is intended for second- and third-year university students, we have tried to make it accessible to the interested reader outside of an academic setting. We have attempted, wherever possible, to provide a clear and basic understanding of the physiological processes being discussed. However, we realize that many readers will not have the background to understand all of the concepts introduced in this book. For this reason we have provided a comprehensive glossary that includes definitions and descriptions of the topics covered.

Our understanding of the impact of stress on reproduction is changing on an almost month-by-month basis. It is not possible to include every theory and advance that has been published. We have tried to provide a foundation for a basic understanding of the effect of stress on reproduction and have introduced new concepts that will probably have a bearing on future studies.

When we first considered writing this book, many of our colleagues encouraged us to discuss the numerous aspects of stress and reproduction across a wide range of all multicellular animals, not to mention those found in fungi and plants. Undoubtedly, the mechanisms of stress on reproduction on invertebrate animals, plants and fungi are very interesting, and in many cases, exotic and unusual by vertebrate standards, but we had to concentrate on a single group of organisms understandable to most readers in order to maintain a focus. Interested readers are encouraged to read and study the mechanisms of stress and reproduction as they will inspire the imagination and study of those biological mechanisms so different from those we typically understand.

We hope that you will find the material covered in this book compelling, but remember that it is only a very small number of species relative to all forms of life on the planet that have been discussed.

Acknowledgements

No book is, of course, the sole result of the authors, and this book is not an exception. Its production is a result of the combined effort of numerous individuals who contributed their time, ideas and resources over the two years we spent writing.

This volume would not have been possible without the support and encouragement of John Wiley and Sons, and their editorial team. In particular, we wish to thank Nicky McGirr who encouraged us to propose the project and was a tireless cheerleader throughout the project, Fiona Woods who kept us organized and on schedule and Celia Carden who handled contract and review details. In addition, we would like to thank Harriet Stewart-Jones, Sarah Karim and Prakash Naorem, for looking after the final edits, galley proof and production details. It is hard to imagine a better team of editors!

The topics covered in this book were the collective result of discussions with numerous colleagues and friends who suggested many of the concepts covered. Professor Robert Dores at the University of Denver provided considerable insight into the evolution of the stress response, Professor Franco Vaccarino and Dr Susan Rotzinger at the University of Toronto spent many hours discussing the relationship of stress with anxiety and depression with us, and Professors Ted Brown and Denise Belsham in the Department of Medicine at the University of Toronto kept us abreast of the latest developments in reproductive physiology. We owe special thanks to Professor Dr Jackson Bittencourt at the University of Sao Paulo and Dr Jean-Michel Aubry at the University of Geneva Medical School for their ideas and understanding of the neurobiology of stress.

We owe a great debt of gratitude to Dr Ian Dunn at the Roslin Institute and Dr Kevin O'Byrne at King's College, London for critically reviewing an earlier draft of this manuscript. Their experience, insight and suggestions had a huge impact on this book.

Completion of this book would not have been possible without the understanding and hard work of the graduate students and research associates in the laboratory: Dr Claudio Casatti, Laura Tan, Dhan Chand, Tiffany Ng, Lifang Song and Tanya Nock. Without them, we could have never found the time to write this book.

And we especially thank John, Natalie, Gillian, Rachel, Paul and Tammy at the 'Harbord House' in Toronto, who kept us plied with nutrients while we composed a significant part of these writings. You guys are the best!

1

Reproduction under safe conditions

> When two great forces oppose each other, the victory will go to the one that knows how to yield.
>
> Lao-Tzu, *Tao te Ching* (sixth century BC)

1.1 Introduction

Most of us regard the act of reproduction as a rather private affair. Despite the volumes of books and magazine articles written on sex and reproduction, and its acceptance into public consciousness, we feel uncomfortable discussing sex with a crowd around us. Sex is intensely personal. It's difficult to be romantic with a partner at midday on a crowded city bus, at a football game or when your children are running around the house. No, we prefer those moments of peace when we are alone with our partners. We might put on some music, turn down the lights and unplug the telephone. We create an environment in which we feel calm, relaxed and safe. We don't think about why we do these things, it is intuitive and natural. And we certainly don't consider the results of a few billion years of evolution encouraging us to reproduce under safe conditions.

Reproduction is the primary goal of all forms of life. Without the ability to reproduce, there is no life. This aspect defines all life forms regardless of whether it is a bacterium, protozoan, a plant, fungus or animal. There are a multitude of strategies various life forms have adopted to ensure they reproduce. These include fission of single-celled organisms, the budding of a smaller individual from a larger individual or the fusion of cells, which, in the case of sexual reproduction, leads to the development of a new individual. In

Sex, Stress and Reproductive Success, First Edition. David A. Lovejoy and Dalia Barsyte.
© 2011 John Wiley & Sons, Ltd. Published 2011 by John Wiley & Sons, Ltd.

some cases, a combination of these strategies may be employed. But regardless of the reproductive strategy used, once the new individual is 'born', it must survive in a hostile and unforgiving environment long enough for it to reach a stage of maturity at which it too can reproduce. So, if the primary goal of a species is to reproduce, then a secondary imperative of the individual is to survive long enough until it can itself reproduce. Because the time required to reach reproductive maturity for all organisms is much longer than the time required for the act of reproduction, organisms have evolved a number of strategies and mechanisms that allow them to survive the conditions of a harsh environment.

The environment surrounding an organism is dangerous and constantly challenges its ability to survive. There are seasonal and daily temperature fluctuations, and an atmosphere that allows organisms to respire but is toxic in some ways. There is ionizing radiation from the sun and cosmos. There are mechanical threats in the form of geologic activity, severe weather, wave action, shifting sand and wind. Food sources may be plentiful at some times, but unavailable at other times. And while you, the organism, is trying to survive, other organisms may be interested in attacking you – either larger predators looking for a meal or much smaller ones which cause a variety of diseases. Added to these stressors are the toxins and noxious chemicals that are found in all environments from a variety of sources. We call these aspects that threaten our survival, 'stress' (Figure 1.1).

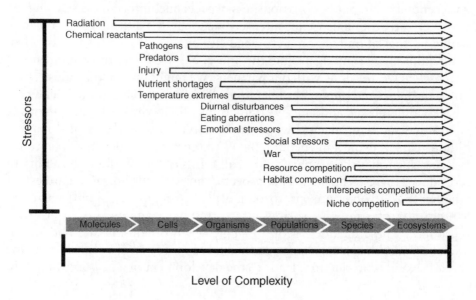

Figure 1.1   Types of stressors that act on various levels of biological complexity. The arrow associated with each stressor indicates the range of complexity that it can affect

One of the fundamental principles of the life history of any organism is that the evolution of fitness-related traits will be constrained by the presence of trade-offs between them. In other words, a beneficial effect on one physiological system can have a negative effect on the expression of another. If we were to consider this with respect to reproduction and stress, an organism could increase its reproductive capacity with a reduced ability to ward off stressful challenges or it could improve its ability to handle stress but with a reduced reproductive capacity. Therefore, for most species there is a compromise that provides a certain level of reproductive capacity with an appropriate level of a stress response that is designed to meet most, but not all, of the challenges for that species.

1.2 What is stress?

Most of us have an intuitive understanding of what constitutes 'stress'. For those of us living in an urbanized civilization, most of what we consider as 'stress', we experience with psychological and social interactions. We face many of the same problems as other animals as well as others that are unique to our species, such as loss of employment, overwork, too many bills and traffic conditions, to name a few. Under normal circumstances, these stressors rarely bother us, but when these events stop us from carrying out our day-to-day activities, we recognize these events as stress. When stress occurs over a long period of time, we might experience anxiety, depression and a variety of other conditions such as post-traumatic stress disorder, panic disorder or various phobias. Then if an additional stressful situation occurs, medical treatment might be required. In our Western style of living, we have a culture based around the stress of living. A quick perusal through newspaper and magazine advertisements will see that the media encourage us to reduce stress by being pampered by spa treatments, take a getaway cruise to a tropical isle or take out golf or gym memberships. The reality is, of course, that we have to indulge in these stress-reducing activities within the confines of the free time allotted to us and within our financial budgets. For the majority of us, both time and money are limited. Excessive indulgence in these commercially related stress-reducing activities can cause debt and reduce our time available for work. This struggle to fit relaxation time into our daily lives might even increase our stress load. We strive for a balance in life, but only a very few actually achieve it.

The concept of stress was originally recognized as a condition associated with humans, but as we came to understand more about the physiology of stress, we could see that it could be applied to all species. What we routinely define as stress, in the biological sense, has a long history. Hans Selye (1907–1982), a

Hungarian physician and scientist, who later emigrated to Canada, coined the term 'stress' to account for a group of similar symptoms in patients who were suffering from seemingly unrelated conditions. He agreed with the findings of the French surgeon, René Lerviche, who in 1934 described a clinical syndrome with similar symptoms which appeared after a wide variety of severe surgical interventions. He called this syndrome *la maladie post-opératoire*. This observation was consistent with a set of characteristic changes following exposure to certain drugs, infections, nervous stimuli, trauma and burns, suggesting that, despite the great variety of noxious insults to the body, there was a characteristic response in each case.

Selye borrowed the term 'stress' from the physical sciences to describe this phenomenon, although in his later writings he suggested that the term 'strain' might have been more appropriate. He speculated that perhaps in all the stresses and strains of life, the body could mount a similar defence mechanism to a specific noxious stimulus. This insight was a major step forward in our understanding of stress. What we understand as stress – a condition that inhibits us from carrying out our normal life – he recognized as the symptoms associated with a defence mechanism that the body elicits in order to combat situations that threaten the health of the individual.

Selye also wrote that some stressful conditions were effective in reducing the severity of other potentially damaging conditions, although he was not the first to recognize this – observations that pain, starvation and fever could actually have curative properties date back to early Greek and Roman times. In other words, some forms of stress could actually improve the health of the organism (Figure 1.2). These observations stimulated further research on 'non-specific therapy' which was hoped could elicit a similar physiological response in the body that would help cure other conditions.

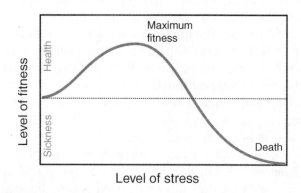

Figure 1.2 Relationship between fitness and stress. Lower levels of stress can increase the fitness of an organism. If the level of stress becomes too high, it becomes damaging to the organism. Fitness is defined as the reproductive capacity of an organism

Intrinsic to an understanding of stress is the concept of 'homeostasis'. The nineteenth-century French physiologist Claude Bernard (1813–1878) recognized the difference between the internal body mechanisms, distinguishing the *milieu interieur* or internal environment from the *milieu exterieur* or external environment, which consisted of all those conditions outside the body that affected an organism's health. The internal environment could be described as the total set of physiological and metabolic reactions that occur within the body. The external environment consists of all processes that occur in nature outside the confines of the body. Bernard wrote that the internal environment had to remain more or less constant despite changes in the external environment in order to maintain life. This 'steady state' of the internal environment was later described as 'homeostasis' by the American physiologist, Walter Cannon (1871–1945). Homeostasis can be defined as the balance achieved by the total set of metabolic and physiological reactions occurring in the organism. Within the context of these definitions, stress was defined as any event that acted to disturb homeostasis, or disrupt the physiological balance of the steady-state internal environment in the body. Biologically, the 'stress response' is the set of all neural and endocrine-associated adaptations that help re-establish homeostasis.

Cannon introduced the term 'fight or flight' to describe the decision the organism makes when confronted by a stressor. In a situation where it is possible to remove oneself from the stressor, such as a predator or temperature extremes, for example, 'flight' ensues. On the other hand, in the case of disease, injury or predation when escape is not an option, 'fight' occurs.

Since its original description, the concept of homeostasis has now been subdivided into two forms: 'reactive' and 'predictive' homeostasis. Reactive homeostasis occurs in direct response to a homeostasis-challenging event (i.e. stress). An example of this might be the sudden threatening appearance of a predator. Predictive homeostasis, on the other hand, is anticipatory, for example when an animal forages for food and anticipates when a food source will become available. The ability to perceive the arrival of a stressor allows an organism to make the necessary physiological adaptation before the arrival of a stressor so that the animal is ready to defend against the stressor. Ultimately, this increases the survivability of the organism. Think how much easier it is if you know ahead of time you have a major examination coming up instead of your professor surprising you with the exam one morning. It is a similar situation with the stress response. Physiologically, it is more efficient to prepare for an upcoming stress than to mount the appropriate defence in direct response to the challenge.

Later, the term 'allostasis' was suggested by Sterling and Eyer (1988) to describe the adaptation of the body to greater challenges such as those brought on by stress. They suggested that an organism might experience greater physiological stability by the exposure to variability in the environment. The greater the range of stressful challenges that an organism can successfully overcome, the greater its ability to confront novel challenges in the future.

Selye speculated that disease results following a prolonged bout of stress because the ability to mount a stress response eventually wanes as a result of depletion of various protective hormones. However, today's evidence indicates, in contrast, that it is over-production of these hormones that produces the symptoms associated with chronic stress. Prolonged activation of the defence systems eventually becomes damaging to the organism. If an organism cannot appropriately initiate a stress response during an acute physical stress, then the consequences can be damaging to the health of the animal. Alternatively, if an organism cannot terminate a stress response at the end of the stress, or if it activates the system too much because of repeated or chronic stressors, stress-related disease may emerge. In other words, a little bit of stress is healthy and too much stress is unhealthy.

In Selye's description of the stress response, he described three stages within the context of what he called the 'General Adaptation Syndrome'. The 'alarm reaction' was the first phase and included the perception of the stressor and the initiation of the stress response. The term 'alarm reaction' was coined to indicate that this response represents a 'call-to-arms', as he described it, of the body's defence forces. Those events that stimulated the stress response were defined as 'alarming stimuli' or stressor agents with an action that was localized and required no important general adaptation adjustments. For example, selectively acting drugs or the amputation of limbs are relatively mild alarming agents even if they maximally stimulate or actually kill limited cell groups. Agents that affect large portions of the body, on the other hand, evoke an intense general adaptation syndrome.

The alarm reaction itself is not necessarily a pathological phenomenon. Stressors might be traumatic to large portions of the body. Here, Selye included haemorrhage, temperature extremes, radiation, electric shock, nervous stimuli including emotions, too much muscular exercise or too much rest, anoxia and asphyxia infection, anaphylactic reactions, general acting drugs and other toxic compounds, diet aberrations such as fasting, overfeeding or dietary deficiencies, diurnal variation or climatic conditions. He also stated that being under or overweight, or having various types of gastrointestinal and menstrual disorders and cardiovascular diseases could be classified as diseases of civilization because they are comparatively uncommon in less urbanized societies. Their development is promoted by food habits and stress inherent in civilized life to which the organismal stress response cannot adequately adjust itself. The level of stress in a technologically advanced urban society may be increasing at a rate faster than that at which humans can adapt genetically to the stressors. Our adaptation to this form of stress comes in the form of behavioural changes and, in some cases, pharmacological interventions.

The second stage of Selye's description of the stress response was the 'stage of resistance' which was characterized by an increased resistance to the particular stressor to which the body has been exposed and, in some cases, associated with a decreased resistance to other stimuli. The final stage, the stage of exhaustion, represented the sum of all non-specific reactions in the body that develop as a result of prolonged overexposure to stimuli to which adaptations have been developed but that can no longer be maintained. All organisms have a limit to how long they can tolerate the stressor.

Robert Sapolsky at Stanford University refined the concept of stress by stating that the original definition of a stress must be expanded to include the psychological aspects of anticipation, rational or otherwise, of physical and emotional stress. The fear and anticipation are particularly important in human society because of the emphasis placed on emotions. The possibility of losing one's financial security or the presence of a bully, even when no action has occurred, is enough to stimulate this stress response. There are examples of students dropping out of school or, in extreme cases, even taking their own life before obtaining their final grades because they were convinced they would not get the marks they wanted, even though their fears were unsubstantiated. In a study involving first-time paratroopers it was shown that the greatest increase in certain stress-related hormones in the blood occurred just before the soldiers jumped from the aircraft. The jump itself caused only a small additional increase in these hormones. There was also shown to be a major increase in some hormones the evening before the jump.

1.3 Reproduction and stress

The physiological processes that regulate reproduction are particularly sensitive to stress. In humans, when sufficiently high levels of stress occur over a prolonged period, women can experience irregular or absent menstrual cycles and men may have reduced sperm counts and impotency. In animals, poor husbandry methods in agricultural or zoo settings can lead to a loss of the ability of animals to mate and breed. It has also been noted that a bout of intense but brief stress can inhibit reproductive processes in the short term, with normal reproductive ability occurring after the stressor has been removed or resolved. In observation of this, Selye wrote 'Curiously, the ovarian atrophy and infertility due to senescence is delayed in mice temporarily kept on a restricted caloric intake. Here one gains the impression that the interruption of sexual life, due to undernourishment, merely "saved" fertility for a later time.'

The longer the time spent to reach reproductive maturity the greater, the chance that an organism will endure a stressful experience by the time it reaches an age at which it can reproduce. Every organism has a unique set of stressors that relate to its niche and habitat. Jellyfish, for example, have to contend with the ingestion of noxious chemicals and bacteria, temperature extremes and predation, whereas for humans these days, the majority of stressors arrive in the form of social and behavioural interactions. Despite these differences, both species must endure and survive these stressors during the passage from birth to reproduction. Thus, there is an optimal window of time in the organism's development where the health and defence ability is maximized to ensure that reproduction can occur. Sapolsky also considered the role of ageing with respect to the stress response. Ageing can be thought of as a time of decreased capacity to respond appropriately to stressors. The aged organism might require less of the stressor for homeostasis to be disrupted or it may take longer for homeostasis to be re-established. In fact, this is likely one of the evolutionary reasons for menopause in women when the reproductive system is shut down when the body can no longer mount the appropriate stress response to deal with stressors. Chronic stress may accelerate some aspects of the ageing effect. Thus, the time at which an organism reproduces has to be optimized for the shortest possible period to reach full reproductive maturity, in which their health is maximal and they are relatively safe from predators and other stressors.

Gestation extends the critical period required to keep stressors at a minimum. The length of gestation varies greatly among organisms. Generally, larger and more complex organisms have longer gestation periods. Also, animals with lower metabolic rates (e.g. shark) tend to have longer gestation periods and those animals with high metabolic rates (e.g. starling) have shorter ones. The gestation period is critical to the development of viable progeny. It is a time when cells are rapidly dividing and differentiating into new cell types. Proliferation and differentiation can be affected by the amount and quality of the nutrients the mother is ingesting, the presence of stress hormones produced by the mother, infection of foreign pathogens or by the presence of toxic chemicals. Prolonged exposure to stressors during this period can lead to a number of developmental abnormalities by altering the physiology of the developing cells, creating progeny that have impaired health and fitness. In general, success of the progeny and, therefore the population, increases if the developmental period is relatively free from these threats. The longer the gestation period the greater the risk that these problems will occur. In many mammals and birds, however, intense stress experienced by the mother may actually modify the physiology of the fetus and embryo so that, as adults, they are better prepared to deal with stressful challenges. These 'epigenetic' effects are discussed in Chapter 7.

1.4   Reproduction, stress and energy are intrinsically interrelated

In the mid 1960s there was an American television series entitled 'Star Trek'. It detailed the adventures of a starship and its crew as they explored the galaxy. What I thought was interesting about the spacecraft was how they used energy. There was a finite amount of energy produced by the spacecraft, which they could use for speed and transport, or protective force fields or weapons but not for all. In many episodes, they had to make a decision about what they were going to use this energy for. I always thought of that spacecraft as a living organism in itself. We might also liken this energy trade-off to your monthly budget: all living expenses need to be paid, then the remainder is divided between saving for the future and spending on entertainment and other interests.

All animals have to deal with the same problems. All organisms have to contend with a finite amount of energy. Organisms obtain their energy through food and convert this energy into useful energy-containing molecules available to the body through the oxidation of nutrients generally with oxygen, although sulphur is used in some chemosynthetic organisms. This need to obtain our food from ingestion of nutrient molecules makes us different from photosynthetic organisms, such as plants, some protozoans and symbiotic animals, and gives us a unique set of stressors that are related to food-gathering.

A finite source of energy forces animals to make decisions about how the energy will be spent. The amount of energy animals can take in is dependent on the amount and quality of food sources available to them, the ability to convert those nutrients into a form that the body can utilize, as well as having the time available to obtain those food sources. An adequate supply and access to nutrients is required for optimal health. However, when food is scarce or time is limited, many animals will opt for a more convenient but less nutritious food source. Stress and energy production are intrinsically intertwined. The hormones associated with stress are also associated with energy production. Activation of the stress response will curtail the need to eat and digest, but may increase the stamina and locomotor ability required for foraging. It may also trigger metamorphosis in some tadpoles whose ponds are in danger of drying up. A similar situation applies during long bouts of migration experienced by, for example, salmon, when the excessive energy demands trigger a sufficient stress response to inhibit foraging. Thus, due to both internal and external stressors, the stress response acts, in part, to maximize energy production to ensure the survival of the organism.

1.5   Interaction of stress and reproduction

The actions of stress on reproduction act at multiple levels of organization in an organism. We would expect, therefore, that the greater the complexity of

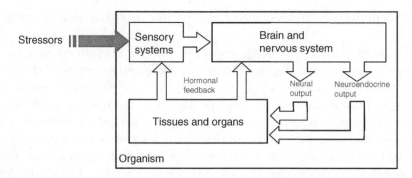

Figure 1.3 Internal effects of stress on a multicellular organism. Stress is perceived by various sensory systems. This information is communicated to the nervous system then out to the tissues and organs by both neural and neuroendocrine signals. Then communicated back to the nervous and sensory systems by secreted hormones

the organism the greater the variety of stressors that would affect it. Let us consider how all of these levels of organization become involved in complex organisms such as humans.

Before a stressful stimulus can affect an organism, the stressful event must be perceived (Figure 1.3). For all stressors in the external environment, activation of at least one of the special senses is required. This may include hearing, vision, touch, taste, smell or balance for most mammals. In bats or whales, it may also arise in the form of echo-location, or in fishes that can perceive electrical signals, changes in an electric field may be sufficient to stimulate a stress response. The strength of this signal is magnified if additional senses perceive a stressor. Once, while I was camping in a remote region of western Canada, I heard the sound of something moving through the forest. Wondering what it might be, I went to investigate and was alarmed to discover that the sound was being made by a large black bear. At this point, it became apparent that both the bear and I were alarmed by the appearance of an intruder, and after appropriately considering 'fight or flight' decision, we both ran in opposite directions. Thus, after the perception of an event in the external environment, and the subsequent identification of the stimulus as a threat, came the associated response. The first sensory response triggered an investigation, whereas the second led to flight, although both perceptions are stressors.

The identification of the threat occurs primarily at the neurological level, although in less complex animals noxious stimuli may be dealt with by reflex action. The processing that occurs in the brain includes a number of associated actions or memories of previous experience or knowledge about the threat. In my case, I knew from experience that the bear represented a distinct threat to my health. Once this occurs, then a number of neural and endocrine mechanisms are activated to help cope with the threat. These mechanisms act primarily to shift the energy reserves to those regions of the body that will be required to deal with the threat. For example, blood flow increases to muscles that will be needed to run or fight, and is shunted away from those systems that will not be needed, such as digestion and reproduction. Sensory systems may also become accentuated to increase awareness. In an acute stress event such as this, the threat passes and soon all systems return to normal. After several minutes of heavy breathing, I calmed down and could enjoy a meal much later that day.

If, however, the stressors continue for a longer period then an additional set of physiological systems are activated that will continue to have a profound effect at the tissue and cell levels of the body. Reproduction, digestion and even the ability to ward off disease can be compromised in the face of continuing threats. This occurs because we are limited in our ability to both engage in growth and reproduction and fight off threats at the same time. It is only after we perceive we are safe from threat that our bodies can return to normal growth, development and reproduction processes. In the case of reproduction, stressors can inhibit our desire and ability to reproduce, affecting the normal development of our progeny and our ability to care for them.

1.5.1 Mechanisms and strategies of reproduction

Organisms have evolved different ways to reproduce. There are two basic strategies of reproduction – sexual and asexual. These have evolved according to the unique collection of selection pressures that have acted on different species. Sexual reproduction involves the mixture of genetic material from two individuals – typically male and female. Asexual reproduction occurs when there is no mixture of genetic material and the offspring are essentially clones of their parent. Examples of this include fission, budding and parthenogenesis. Asexual reproduction in the form of fission is typical among single-celled organisms. Budding may occur in some single-celled life forms, for example yeast, and in simpler metazoans such as hydra. Parthenogenesis, in which there is no fertilization of the egg by males, occurs in numerous species of animal including the more complex fish, reptilian and avian species. It is relatively rare and is generally confined to evolutionarily younger lineages of animals who are living in a relatively unstable environment or when the prospects of finding a mate are reduced. About 70 species of vertebrates have been found to use parthenogenesis as a reproductive strategy. The advantages of asexual reproduction are that it allows for the rapid recovery of a population in the case of a catastrophe, although this strategy ultimately reduces the species' adaptive ability over a longer period of time.

The majority of multicellular animals reproduce sexually. Sexual reproduction mixes genes. Different combinations of genes from the parents lead to numerous variations inherited by the progeny. The widespread occurrence of sexual reproduction was, therefore, more successful than other strategies. However, sexual reproduction is energetically costly and requires more nutrients than asexual modes of reproduction. Parthenogenic and asexual species have much smaller gene pools than sexual species because they are essentially clones of their parent and possess the same complement of genes. As a result they would be expected to have more limited niche adaptation abilities than sexual species. Asexual reproduction also reduces sibling competition because they are genetically similar and will tend to respond to a given stimulus in a similar manner. Thus, in the short term, asexual species may be expected to dominate because of their greater reproductive capacity but over longer periods sexual species would dominate as the competition becomes greater because of greater variability in the gene pool.

Sibling rivalry is an essential aspect of development because it prepares the individual for competition in adult life. Thus, early in the origin of a species, genomic variability may be selected for. For sexual species, the chances of survival increase because the population, in general, has less susceptibility to disease or predation due to the genetic and phenotypic variability of the population.

Sexual reproduction also helps to remove harmful mutations. In asexual species, the number of mutations can only increase with successive generations, because the progeny have all inherited the same genetic material. In species that have a short generation time, changes in DNA due to the accrual of mutations becomes the primary method by which the DNA can change over time, allowing the organism to adapt to changes in its environment or to expand its niche and habitat. In sexual species, however, the offspring may have lower mutations than parents due to the genetic recombination. The trade-off is that whereas the populations of sexual species may be genetically more stable, they may be more sensitive to sudden changes in the environment.

1.6 Evolution of germ cells

One of the requirements of sexual reproduction is the ability for cells from two different parents to fuse and become a single cell. If non-sex cells did this the progenitor cell would have twice as much DNA as the parent and the amount of DNA would increase each time the cells fused with each subsequent generation. But there are limits to the amount of DNA that can be contained within the cell. So, before sex cells, or gametes, could evolve, a method to reduce the total amount of DNA had to be developed. We might imagine that the first sex cells were similar to other cells and the resulting progenitor cell possessed more DNA than the parents. This might have been tolerated for a number of generations until the increased amount of DNA no longer produced an organism with improved fitness. If a process evolved that reduced the amount of DNA and increased the fitness of the cell, it may have been retained by subsequent generations. Such a process may have eventually led to the mechanism we understand as meiosis. Although the origins of meiosis are unknown, it is this process that allows the complement of sex cell DNA to be reduced by half, forming what is referred to as the 'haploid' cell. The diploid cell, on the other hand, possesses a set of DNA from both parents. Originally the sexes were basically similar but each had two types of cells: diploid somatic cells and haploid sex cells. Fusion of the sex cell produced a diploid cell.

The events leading to the evolution of male and female sex cells are not clear. However we might consider one scenario. In the most ancestral sexual organisms the structures of the gametes were probably similar in shape and size. However, over time, two size morphs would have been selected for. The size difference would promote different selection pressures acting upon those cells. Larger gametes would have better survival ability in that they would possess a larger store of nutrients. Thus, in this cell type, acting to maintain this morphology, the genes would be selected for. Our understanding of the selection of genes has changed over the years. Richard Dawkins of Oxford University in his book *The Selfish Gene* has been particularly influential in this respect. Although the genetic and molecular mechanisms of gene selection and survival are beyond the scope of this book, this theme will be continued throughout the chapters. Ultimately, the gene needs to replicate, and it might be argued that cells and all organisms, in general, have evolved to efficiently proliferate particular genes.

At the other extreme, smaller sizes may have had the advantage that they could be produced more quickly with fewer nutrients. Their long-term survivability would not be that important because they could fuse with the larger gametes. Intermediate-sized gametes would have no selection advantage and would eventually be eliminated. If so, then large gametes become the female form, whereas small gametes would become the male form. As a result, females become selected for the ability to produce larger but fewer gametes. In females, physiological systems aimed at increasing nutrients of eggs and their protection are selected to maximize the survival of the gametes. On the other hand, males become selected for a reproductive system that produces larger numbers of small gametes. As these organisms evolve into more complex forms, the structure of the reproductive system, the metabolic, endocrine and neural pathways and associated behaviour evolve into distinct male and female forms to promote the success of their gametes in the population. We term these physiological and morphological differences between the sexes 'sexual dimorphism'. Because the reproductive requirements of the two sexes are different, the actions of stressors affect them differently.

1.7   Variations in reproductive strategies

One of the difficulties in sexual reproduction can be finding a compatible mate. This is easy enough if the population is large enough and the environment is sufficiently stable. However, for many species the population may undergo drastic decreases due to predation, disease or climatic and geological catastrophes, for example. Following such events, the chances of finding a mate are reduced. In addition, an unequal number of males or females may be lost, causing the population to be skewed toward too many females or males. The loss of females in particular reduces the number of progeny and puts stress on the continuation of the population. Ultimately, this reduces the gene pool of the population and, therefore, the fitness.

A number of strategies have evolved to solve this problem. One way to ensure that you can mate when you encounter an individual from the same species is to change sex to the one opposite your potential mate. Such 'sex reversal' occurs naturally in numerous species of fish as well as a number of invertebrate species. If every member of a population has the potential to mate with any other member of the population, then this doubles the gene pool size. There is a downside to this phenomenon. Many of the pollutants released into the environment by human industrial activity have actions that are similar to the reproductive hormones determining sex. This has led to complete or partial sex reversal in sensitive species – mostly males to females. As a result, it has reduced the number of viable males in the population, causing a reduction in the gene pool. In addition, the sex reversal process introduces a new developmental window into the life of the organism when it may become more sensitive to environmental stressors. Again, sex reversal is an energetically costly affair, and it has only evolved in organisms in which the advantages offset the disadvantages produced by the extra energy demand and the sum of other external stressors.

Another evolutionary strategy for sex determination is temperature-dependent determination. A number of reptiles, including all crocodilians, some lizards and tortoises utilize this approach. Here, the embryo develops as male or female depending on the ambient temperature. Although animals using temperature-dependent sex determination could risk loss of fitness due to a skewed sex ratio if males were favoured in response to sustained environmental temperature changes, for example, they have the advantage that they can capitalize on the gene pool of the entire population. Another advantage is that they do not possess the potential problem of the evolutionary decay of the male sex-determining chromosome, due to non-recombination and the subsequent loss of spermatogenesis genes. However, temperature-dependent sex selection may involve a number of both environmental and genetic factors. Thus significant changes in global temperatures could have an impact on such species. Global temperature change may skew the sex ratio of temperature-dependent sex determination animals and might have played a role in the extinction of some species, for example, dinosaurs, if the temperature change resulted in an excess of males. If the current global warming trend continues, this might impose a risk for living species that use this mode of sex determination.

1.8   Evolution and complexity

You may have noticed our allusion to the concept that the more complex an organism, the more complex the interaction between stress and reproduction. But what does this complexity refer to? We have an intuitive understanding of this. A snail is more complex than an amoeba, a fish is more complex than a snail, and a mouse is more complex than a fish. Note that size, or the number of cells in the organism does not necessarily indicate complexity. When we describe complexity from a biological point of view we need to be more pragmatic. Ultimately, biological complexity is essentially related to the number of unique functional genes and their promoters. The mass, and therefore, the number of cells in an organism is not necessarily a reflection of complexity. Rather, it is how those cells are organized, how they communicate and how many different cell types are present.

The first organisms that evolved were simple one-celled creatures. Complexity increased over evolutionary history. But how does this happen? The processes of meiosis and mitosis are not perfect. In addition, the mechanisms of DNA breakage and repair and chromosomal duplication are also not perfect, and periodically genes, chromosomes and even the entire complement of genetic material (genome) duplicates. Sometimes when this happens a survival advantage is conferred on the organism. The more proteins a genome produces and the greater the number of different ways the expression of these proteins can be regulated the greater the number of ways the environment can interact with the organism. When genes or genomes duplicate, the organism takes on a second set of identical genes. But it only needs one set to survive. The theory is that if the first gene is important to the survival of the organism then the second gene is free to accrue mutations. If these mutations lead to decreased reproductive fitness, then the organism or its lineage eventually dies out. If, however, these mutations lead to an improved function of the gene which also improves the reproductive fitness, then the gene and the organism it is found in will survive. Thus, as genetic complexity increases, the niche complexity that an organism can adapt to increases. Therefore, animals with more complex genomes become selected for. For stress and reproduction, this means that the more complex an organism, the greater the number of genes that can ultimately respond and deal with different threats and stresses.

1.9   Summary

Organisms must survive and reproduce in a hostile environment. The events in the environment that act to disrupt the health of an individual are collectively referred to as 'stress'. Stress, whether real or perceived, affects the homeostasis of organisms. A small amount of stress can be beneficial to the organism, but if it becomes too great, it can offset the homeostatic balance and the organism's health becomes compromised. Different stressors act on all levels of complexity in an organism. The ability of an organism to cope with these stressors is referred to as the 'stress response'. The stress response begins as a set of neural and endocrine changes that act to regulate the metabolism and behaviour to fight or avoid the stress as much as possible. Because reproduction is essential to the survival of the species, mechanisms have evolved to ensure that reproduction occurs in a stress-free environment as much as possible. People living in Western urbanized societies are facing increased amounts of stress because the rate of stress is outstripping our ability to adapt. As a result we are seeing a higher incidence of reproduction and sex-related problems.

2

Reproductive physiology: how is it all supposed to work together?

'He does not love me for my birth, Nor for my lands so broad and fair; He loves me for my own true worth, And that is well,' said Lady Clare.

Alfred Lord Tennyson, 'Lady Clare'

2.1 Introduction

Up until relatively recently, in historical terms, people believed that simple living organisms came into existence spontaneously. In the seventeenth century, for example, it was thought that mice could be created by simply putting a dirty shirt with some grains of wheat together in a jar, and the Royal Society of London was seriously discussing how to generate vipers from dust. Because the rules of reproduction and generation were not understood until the mid-nineteenth century, there was still discussion of how species could interbreed and produce chimaeric monsters showing traits of both species. Such concepts survive even today in some parts of the world. When I was undergraduate student back in the 1980s, the manager of my apartment building told the story of how a cat and dog had mated to produce animals that were cats in front and dogs behind. I challenged her on this, saying that such things were impossible, but she continued to insist the story was true. In the true spirit of tact and diplomacy, and needing to stay on the good side of the manager, I did not pursue the argument. But what I realized is that in situations like this where there is no knowledge or understanding of the mechanism of reproduction and genetics, then all interpretations have equal weight.

Sex, Stress and Reproductive Success, First Edition. David A. Lovejoy and Dalia Barsyte.
© 2011 John Wiley & Sons, Ltd. Published 2011 by John Wiley & Sons, Ltd.

In the not too distant past, the relationship between sexual intercourse and pregnancy was difficult to establish. Cessation of menses was variable and the honesty of intercourse was an issue. Thus, without a theory it was difficult to establish the mechanics of cause and effect. In the early eighteenth century, medical texts provided the first reasonably accurate accounts of embryonic development and attempted to explain the relative contributions of egg and sperm. However, it was really only in the twentieth century that conception and contraception were understood. Thus, to appreciate how stress can affect reproduction and sex, it is useful to have a basic understanding of the mechanism of reproductive physiology and behaviour.

In the last chapter we discussed how reproduction must be successful in a hostile environment. The methods that chordates (animals with backbones) have used to achieve reproduction are numerous, and it is fair to say that the mechanism of reproduction differs in some detail in every single species. Each unique system has evolved in response to the constraints imposed upon it by its niche, habitat and internal physiological, biochemical and genetic attributes. Thus, many of the stressors will be unique to a given species or group of related species. Despite this, there are a number of common elements in all chordate species.

Depending on the type of stressor, its primary impact on the organism can occur at different levels of complexity as discussed in the previous chapter. In the specific example of reproduction, this means that specific stressors can act at the cellular level to affect the proliferation, migration and differentiation of the gametes, or at the organ level to alter the function of systems associated with the maintenance, maturation and conveyance of the gametes and zygotes. Stressors can also affect the neurological processing of the reproductive system, such that the neuroendocrine cycles are compromised, causing disruption or cessation of the coordination between courtship and sexual behaviours with the proper functioning of the reproductive system.

2.1.1   Components and development of the reproductive system

The reproductive system can be thought of as having four basic components. The simplest level comprises the haploid sex cells or gametes. These are the cells that are required for the development of progeny. In chordates, which are the focus of this book, the gametes are housed in a specialized organ system devoted to the proliferation, health and maturation of the gametes. This is the ovary in females and the testes in males. The second component includes the duct system. A set of ducts, specialized for either egg or sperm transport, has evolved to convey the gametes from their organ of production to a region where they can be fertilized and extruded to the exterior. In both sexes there are typically a number of modifications of these ducts to maximize the efficiency of fertilization, maturation and development of the embryo.

Primitive chordates, such as hagfishes and lampreys, however, do not possess such duct systems. In these species, the gametes are initially released into the body (coelomic) cavity, where they are extruded into the external environment through a genital pore at the appropriate time.

A third element of the reproductive system may be thought of as the opening of the duct system to the external environment. It may be something as simple as a genital pore, or consist of elaborate copulatory and vaginal structures found in more complex chordates. The final component of the reproductive system is the nervous system. Although we generally do not think of the nervous system as part of the reproductive system, there are sets of neurological structures, nerves and neuroendocrine organs responsible for the coordination of the reproductive system activity, sexual behaviour, mate recognition and appropriate environmental conditions which are required for successful reproduction.

Before we consider the mechanisms of reproduction, it is useful to understand how the system develops. In the early stages of reproductive development, the pattern of development is similar for both sexes, and as we will see in later in Chapter 7, the development of the reproductive tissues and sex cells plays a major role in the interaction between stress and reproduction at later stages in life.

Although we discussed in Chapter 1 that sex cells are haploid and somatic cells are diploid, the sex cells develop from somatic cells of the embryo. Early in embryogenesis, the genital ridges, an out-pocketing of tissue, develop in the coelomic cavity from mesoderm next to the primitive kidney, the nephrotome (Figure 2.1). The primitive sex cells, derived from early stage somatic cells, migrate from the gut wall into these ridges. As they go through this migration, they pass by other cell types and are affected by the various gradients of secreted materials from the surrounding cells. Once this occurs, the primitive gonad is ready to differentiate into either the testes or ovary. At this stage, the primitive gonad is neither male nor female in function or morphology and the primitive sex cells are found scattered throughout the structure.

Of the three basic cell types found in the undifferentiated gonad, the primordial sex cells produce gametes, although it is not clear if this is the situation in all classes of chordates. The interaction of the primitive sex cells with the other cell types will determine their differentiation into either ova or sperm. One cell type is associated with the epithelial layer of the coelomic cavity which becomes the outer layer of the undifferentiated gonad. The second set of cells is derived from nephron cells that have de-differentiated into a less specialized cell. When the primordial sex cells begin to migrate into the

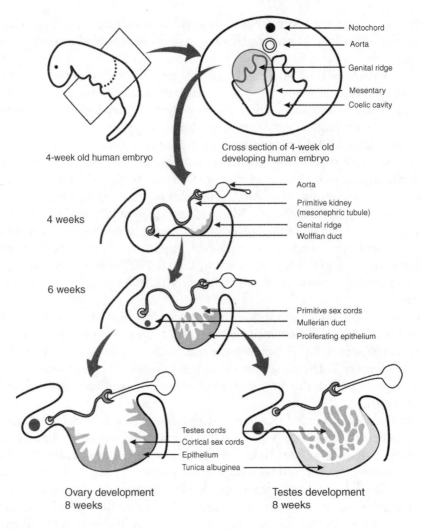

Figure 2.1   Embryonic gonadal differentiation in humans

epithelial layer of the primitive gonad, the epithelial cells and sex cells begin to proliferate and differentiate. As a result, this outer layer thickens and becomes the cortex, or outer layer. The primordial sex cells present in the internal layer along with the de-differentiated nephron cells also proliferate to form the medulla (middle) of the gonad. The cortical cells attract additional migrating sex cells but inhibit cell division, whereas the cells of the medulla stimulate the migration and proliferation of the primordial sex cells.

In females the cortical layer dominates, forming the ovary, whereas in males the medullar layer dominates, forming the testes. During the development of the primitive gonad two duct systems develop: the Wolffian duct, which is associated directly with the undifferentiated gonad, and the Mullerian duct system, which links the coelomic cavity with the cloacal opening. Thus, for a brief time, the gonad remains sexually undifferentiated, having attributes of both testes and ovaries. This same process will occur regardless of the genetic sex of the individual. For this reason, interference with this developmental stage by a variety of stressors, such as xenobiotics and other chemicals in particular, can alter the gonadal differentiation pattern to lead to full or partial sex reversal, which is frequently associated with reduced reproductive potential and sterility.

Sex determination among vertebrates is highly varied and can be fairly complex. In general, under normal circumstances either the testes or ovary will develop from the undifferentiated gonad, according to the genetic programming of the individual. In chordates there are a couple of basic types of sex-determination genetics. The more familiar is the XY/XX mechanism, as is typically found in mammals. Here, males are referred to as the heterogametic sex because they possess two different sex chromosomes. Females, possessing two of the same type of chromosome, are called the homogametic sex. This mechanism of sex determination is typical of most mammals. The Y chromosome contains the sex-determining gene, SRY, which when present will direct the undifferentiated gonad to develop into testes.

But this is not the only system. In birds, for example, females are the heterogametic sex and males the homogametic sex. This mechanism is defined as the ZW/ZZ system. Until recently it was thought that the W chromosome was dominant and directs the undifferentiated gonad to develop as an ovary, however, recent studies have indicated that the Z chromosome may also play a role in the direct triggering of male development.

In many reptiles and some amphibians, sex determination occurs on the non-sex chromosomes, the autosomes, and is associated with ambient temperature. Some reptiles have been found to use both temperature-dependent and XY/XX methods of sex determination. In fishes there is considerable variability as both XY/XX and ZW/ZZ systems have been found, sometimes within the same genus. The sex-determining systems in fishes, however, are different from the XY/XX and ZW/ZZ systems found in mammals and birds, respectively. In amphibians, it has been suggested that in an ancestral period females may have been the heterogametic sex and that later, through some evolutionary 'bottleneck' period, amphibians converted to the XY/XX system, with males becoming the heterogametic sex. Today, both methods of sex determination can be found in amphibians.

Examination of a number of sex-determination mechanism in chordates suggests that transformations between temperature-dependent sex determination (TSD) and genotypic sex determination (GSD), on the one hand, and

XY/XX and ZW/ZZ mechanisms, on the other hand, have occurred numerous times in the past. The reasons are not clear, but do suggest that the mechanism of sex determination in chordates is relatively plastic and has been modified in the past in response to internal and environmental constraints upon the organism. One possible reason for this transformation is that it would protect the heterogametic chromosome, either Y or W, from eventual decay due to the accrual of mutations over a long period. In fact, some authors have argued that mammalian males may not endure because of this eventual decay of the Y chromosome, although this is not likely because of the alternative genetics of sexual differentiation and the ability of lineages to undergo these transformations. Nevertheless, it is not fully understood what triggers such transformations.

Regardless of the method of sex determination utilized, once the program is initiated, there will be a cascade of genes that lead to the differentiation of the sex. In most male mammals, at an appropriate stage in development and after the beginning of the differentiation of the gametes in the medulla of the indifferent gonad, the sex-determining region of the Y chromosome in embryonic germ cells is activated to produce the testis-determining factor (TDF), a transcription factor that initiates the production of multiple proteins required for the differentiation of the gonad into the testes. Among these new cell types will be Leydig cells, which once formed will secrete the steroid hormone testosterone and its derivative, dihydroxytestosterone (DHT), into the surrounding cells and tissues. Both hormones bind to and activate the testosterone receptor, which is actually a transcription factor, to promote further development of the Wolffian duct system into the accessory structures (the epididymis, vas deferens and seminal vesicle) and the development of the male external genitalia. At this stage the Sertoli cells secrete Mullerian-inhibiting substance (MIS), a peptide hormone which promotes the regression of the Mullerian duct.

In mammals, the differentiation of testes and associated reproductive tissues in males occurs before the differentiation of the gonad into the ovary in females. Once differentiation proceeds in a male pattern, natural sex reversal is not possible. In birds and other species using the ZW/ZZ sex-determining system, it is the female reproductive system that develops first in response to the initiation sequence, and the male develops in the absence of this system.

This mechanism contrasts with that in teleost fishes, in which the gonad remains partially in an undifferentiated state. In many species of fish, males can change sex to female and vice versa as a normal part of their life history strategy. In the female embryo, with the absence of the TDF cascade and resulting androgenic hormones (testosterone and DHT), the medulla of the indifferent gonad regresses and the cortex expands, driving the differentiation of the indifferent gonad into an ovary. At the same time, because of the absence of MIS, the Mullerian ducts persist and develop into the upper portion of the vagina, the uterus and oviducts (Fallopian tubes). The external genitalia take on a female morphology. The Wolffian ducts regress in the absence of testosterone and DHT.

Interestingly, mammals may share the same genes associated with sex reversal found in fishes, although they are probably under different regulatory controls. A gene called FOXL2, found in both male and female mammals, interacts with another gene, SOX9, which is found only in males. During male development, FOXL2 is activated first and inhibits SOX9. But if FOXL2 is artificially inhibited in adult female mice, then the ova die and some of the cells in the ovary begin to differentiate into Sertoli-like cells and start to produce testosterone. It is anticipated that turning off the SOX9 gene in males may stimulate the formation of thecal- and granulosa-like cells, leading to the production of female sex steroids.

The embryological development of the gonads, ducts and external genitalia can be thought of as the first of three critical stages of development of the reproductive system. In mammals, a second period occurs around the time of birth, parturition. Evidence suggests that this period is responsible for differentiation of parts of the brain into the male or female pattern. In males, plasma testosterone levels increase dramatically at this time and act on specific regions of the brain to modify the neuronal circuitry for the male reproductive system, copulatory organs and neuroendocrine cycles. This differentiation, along with the testosterone surge, is implicated in the development of normal sexual behaviour and sexual orientation in male mammals. Many of these studies remain controversial to a degree, and it is not clear how much these mechanisms relate to humans. Paradoxically, in some parts of the brain, testosterone is first converted into oestradiol which then acts downstream to initiate the male pattern. In other parts of the brain, testosterone acts directly on the neurons. In female mammals, the female pattern of the brain develops in the absence of testosterone after the critical masculinizing period has passed. The more complex the reproductive cycles and behaviour in a species, the more critical this neurological stage of development will be.

The third critical stage of reproductive development is sexual maturation or puberty. Puberty is a time of development from the juvenile form to the adult form. An adult, in the general sense, is an individual who is reproductively functional. During this time, the adult reproductive cycles begin and the resulting increased concentrations of testosterone in males or oestradiol and progesterone in females act to stimulate the development of the secondary sex characteristics and organize neural circuitry to carry out the appropriate motor and sensory programs required for sexual and reproductive behaviour.

In birds, who also possess relatively complex reproductive physiology and behaviour, this neurological development appears to be more associated with the final critical stage of reproductive and sexual development, in other words, puberty. In many bird species, regions of the brain associated with male singing during courtship may become enlarged at the beginning of each season and regress at the end of the season.

These three critical stages of gonadal, neurological and secondary sex characteristic development are particularly sensitive to the actions of stressors.

2.2   Neurological regulation of reproduction

In its most basic sense, the brain is simply an organ that coordinates the information from various sensory systems to provide an appropriate motor response. For example: 'I perceive food. Must go and eat that food' or 'I perceive a potential mate. Must go and have sex with that mate'. As animals become more complex, the brain still has the same basic function, but more regions are devoted to the encoding of memories, the recognition of complex sensory patterns and the coordination of more complex motor patterns. As the brain continues to evolve, then self-awareness and the concept of spirituality develops and then we develop the ability to consciously modify the needs that our body is telling us.

Although not all chordates will integrate the same kind of information, they will integrate several internal and external variables to elicit an appropriate response. Remember that the more complex the organism, the greater the number of variables that are processed and eventually integrated into an appropriate response.

One of the more primitive elements of the chordate brain, and one that is responsible for the primary integration of stress, sex and reproduction, is the limbic system. The limbic system consists of a number of specialized regions found in the most anterior region (or superior, in the case of humans and bipedal primates) of the brain, called the forebrain. The forebrain consists of two major regions, the telecephalon and the diencephalon (Figure 2.2). The term 'limbic system' was first used by Paul MacLean in 1952 and encompasses the limbic lobe, which was originally introduced by Paul Broca in 1878 to describe the hippocampal formation and cingulate gyrus, the parahippocampal gyrus and other structures in the brain, such as the amygdala, septal nuclei and related parts of the striatum and diencephalon, that are well connected to the hippocampal formation. In mammals, behavioural studies indicate that the limbic system is involved in motivation, memory, learning, fear, sexual and aggressive behaviours, and plays a significant role in the integration of the

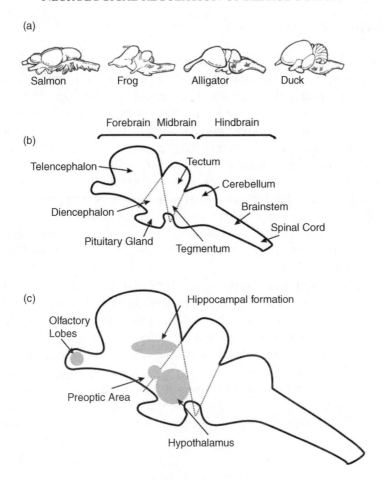

Figure 2.2 General organization of the vertebrate brain. (a) Representative brains from various species. (b) Main regions of the brain. (c) Key regions discussed in this chapter

sensory input to provide the required output responses. The septal nuclei and amygdala receive direct input connections from the olfactory system and indirect input from the other sensory systems. The limbic system is also intrinsically associated with the regions of the brain associated with the perception and integration of the stress response. This will be discussed in greater detail in the next chapter.

Although most of our understanding of the limbic system has been gleaned from studies on the mammalian brain, the similarity with many of the components in non-mammalian species suggests that the system works in a similar manner, albeit with less complexity, in sensory perception, integration or physiological and behavioural responses.

2.3   Reproductive cycles

There are two or three main components to the regulation of reproductive cycles, depending upon the complexity of the organism. Initially, a neuroendocrine factor is secreted from the brain in response to the appropriate stimuli to initiate the reproductive cycle. This factor then may stimulate the gonads directly, or act via a secondary endocrine organ to regulate the gonads. The neuroendocrine factors are typically neuropeptides. These factors act on the gonads, which, in turn, release a variety of steroid hormones. These gonadal steroids then regulate the synthesis and release of the neuropeptides by a steroid feedback mechanism. In vertebrates, there is a three-tiered system consisting of a neuroendocrine factor or a gonadotrophin-releasing hormone in the brain that stimulates the gonadotrophins from the pituitary gland, which subsequently stimulate the gonads to synthesize and release steroid hormones (Figure 2.3).

The pituitary gland, which is found in all chordates, albeit with very different morphologies, has been considered the 'master gland' by historical researchers. The mechanism by which the brain, and particularly the hypothalamus, regulates the pituitary was of considerable interest in the early part of the twentieth century. The pituitary gland, also termed the hypophysis, consists of two main regions, an anterior pituitary lobe, or adenohypophysis, and a posterior lobe called the neurohypophysis (Figure 2.4). In some chordates

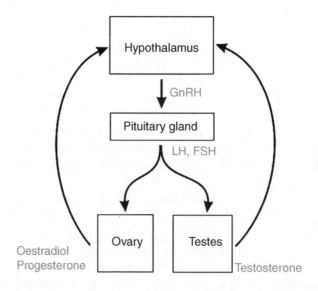

Figure 2.3   Basic reproductive cycle. The main hormones are indicated in grey

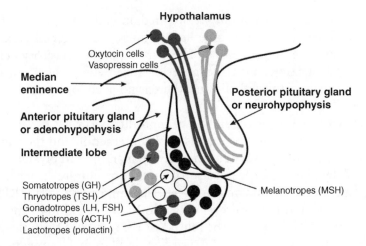

Hypothalamus

Oxytocin cells
Vasopressin cells

Median eminence

Anterior pituitary gland or adenohypophysis

Intermediate lobe

Somatotropes (GH)
Thryotropes (TSH)
Gonadotropes (LH, FSH)
Coriticotropes (ACTH)
Lactotropes (prolactin)

Posterior pituitary gland or neurohypophysis

Melanotropes (MSH)

Figure 2.4 Schematic drawing of a generalized mammal pituitary gland. Note that the anterior lobe contains cells with the hormones, whereas the posterior lobe contains nerve endings containing the hormones

there is a third region called the intermediate lobe, which is found between the other two lobes.

The anterior lobe contains a number of different cell types that produce hormones, such as growth hormone, thyroid-stimulating hormone, adreno-corticotrophic hormone and prolactin, which interact with the rest of the body. It also secretes the two reproductive hormones, called the gonadotro-phins, namely, luteinizing hormone (LH) and follicle-stimulating hormone (FSH).

The neural lobe of the pituitary gland has a somewhat different morphology. Here, the hormones oxytocin and vasopressin (or vasotocin as found in a number of chordates) are stored. They are actually produced in the hypothal-amus and are transported via axons into secretory nerve terminals in the posterior lobe. This means that the posterior lobe hormones can be released directly by other chemical signals acting directly on the hypothalamic neurons. However, because the cells of the anterior pituitary gland are not connected directly to the rest of the brain by a nervous connection, a neurohormone signal from the hypothalamus must be first produced and released to travel through the blood vessels into the anterior pituitary gland. A small swelling in the middle of the base of the brain, called the median eminence, is the site where these chemical signals from the brain are released into a capillary network called the pituitary portal system. Once these chemicals have entered the portal system bloodstream, they are carried to the cells of the anterior pituitary gland.

In fishes, and in particular, teleost fishes, the structure of this system is different, although it has the same effect. Instead of the hypothalamic neurosecretory cell axons ending at the median eminence, they travel into the posterior lobe. The neural lobe has a number of finger-like extensions that project into the anterior lobe. The hypothalamic secretory cell axons follow these projections into the anterior lobe where they can connect directly to the cells of that lobe. Thus, there is no need for a portal system in most fishes.

The basic scheme for the regulation of reproductive cycles in vertebrates is largely conserved, although the details of the interaction and the subsequent timing between events can vary enormously. In mammals, reproductive cycles are initiated by the pulsatile release of gonadotrophin-releasing hormone (GnRH), also called luteinizing hormone-releasing hormone (LHRH), from the hypothalamic and forebrain regions into the portal blood supply of the median eminence. This induces the release of FSH and LH from the gonado-trophs of the pituitary gland into the systemic circulation. Despite the name of these hormones reflecting the function of the female gonad, the hormones are identical in males and females. In males, FSH acts to stimulate the activity of Sertoli cells in the seminiferous tubules of the testes to aid in spermatogenesis. LH, on the other hand, stimulates the production of testosterone from the Leydig cells of the interstitial region of the testes. Most testosterone secreted from Leydig cells passes into the bloodstream. However, a small amount passes into the seminiferous tubules. Testosterone, being a steroid and highly fat-soluble (lipophilic), passes freely across the blood–testes barrier. Sertoli cells secrete androgen-binding protein to maintain local high concentrations of testosterone to ensure that testosterone is present. Testosterone circulates in plasma bound to sex steroid-binding proteins or other plasma proteins.

Depending on the tissue, testosterone can be converted to dihydroxytes-tosterone, 5α-androstenedione or oestradiol. We generally think of testos-terone as being the only androgen, but there are actually several. The major targets of testosterone include the accessory organs of the male reproductive tract such as the prostate, seminal vesicles and epididymus. Testosterone also has an effect on a number of non-reproductive tissues such as the liver, heart, skin, skeletal, muscle, bone and brain. When plasma testosterone levels rise too high, then they inhibit the release of the pituitary gonadotrophins, and also the release of GnRH from the hypothalamus. Once GnRH secretion falls, testosterone levels fall and GnRH is secreted again. This feedback mechanism is on-going, such that testosterone production and spermatogen-esis remains more or less consistent during the lifetime of a non-seasonal breeding male mammal.

In females, the reproductive cycle is more complex (Figure 2.5). As in males, GnRH initiates the reproductive cycle, but in females its pulsatile character-istics are modulated to focus the stimulation on FSH or LH release. During fetal life, the primordial germ cells of the ovary form and continue mitotic

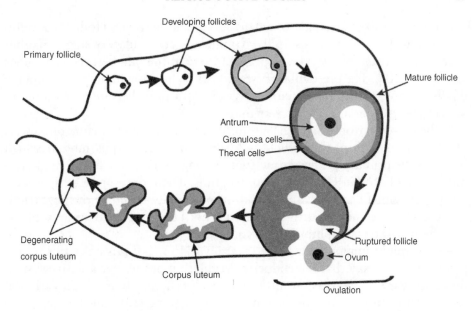

Developing follicles

Primary follicle

Mature follicle

Antrum
Granulosa cells
Thecal cells

Degenerating
corpus luteum

Ruptured follicle

Ovum

Corpus luteum

Ovulation

Figure 2.5 Maturation of the ova in a generalized mammalian ovary

proliferation. By the time of birth, the total number of primordial germ cells is complete. In the fetus, the oogonia enter the first meiotic division and become oocytes. Oocytes are surrounded by mesenchyme cells and a basement membrane (basal lamina) is formed. This is then called a primordial follicle. Oocytes in the primordial follicle remain arrested at the first meiotic phase until reproductive maturity (puberty). Towards the end of the preantral stage, the granulosa cells develop receptors for oestrogens and FSH. The thecal cells, on the other hand, develop receptors for LH. Both gonadotrophins are required for the preantral follicle to develop into an antral follicle. In response to the gonadotrophins, the thecal and granulosa layers proliferate and begin to differentiate. Granulosa cells secrete fluid around the oocyte. This fluid-filled region is referred to as the antrum.

In antral follicles, where the thecal cells are under the control of LH, the internal thecal layer (theca interna) begins to secrete testosterone, androstenedione and small amounts of oestrogens. The androgens from thecal cells diffuse to the granulosa cells that convert these hormones into oestrogens. Much of the oestrogens diffuse into the bloodstream and act to further stimulate the production and release of GnRH and gonadotrophins. Granulosa cells are under the control of FSH. Thus, the granulosa cells are analogous to the Sertoli cells in the male, and the thecal cells are analogous to the Leydig cells. In the pre-ovulatory follicle, LH receptors appear in granulosa cells in preparation for ovulation.

In humans and other single-ovum ovulating species, several follicles develop to this stage but only one is selected to enter the pre-ovulatory stage. In other species numerous follicles develop. Regardless of the number, the remainder atrophy and die. The pre-ovulatory stage lasts 36 hours in humans. The oocyte finally completes the first meiotic division and now becomes a secondary oocyte. The secondary oocyte is ovulated.

In the pre-ovulatory stage, the granulosa cells begin synthesizing progesterone. After ovulation, the follicle is formed into the corpus luteum which becomes the primary progesterone-synthesizing tissue in the body. Progesterone inhibits both GnRH and gonadotrophins. Then, once primed by progesterone, the GnRH neurons and gonadotrophs are inhibited by the oestrogens as well. This then acts to turn off the cycle until the corpus luteum atrophies, and the progesterone inhibition is removed.

As one might imagine, there are several variations of this mechanism. It is probably fair to say that the endocrine regulation of the reproductive system differs to varying degrees within every species of vertebrate. This variation relates to the morphology of the reproductive system, seasonal changes, mate selection, niche and habitat, to name a few. Although the same hormones and basic reproductive structures are generally present, the timing, sensitivity and coordination of these hormones can differ considerably. Among birds, for example, the corpus luteum does not continue to exist for long following ovulation, a trait that is associated with the absence of viviparity. In all birds examined, the remaining follicle consists primarily of granulosa cells containing progesterone.

However, irrespective of the pattern of reproductive hormone production within a particular species, it is critical that the timing occurs in a consistent manner. Disruption of timing by a number of factors, for example, weight loss or gain, nutrient availability, toxins, infections, lack of sleep or overexercise, can lead to loss of reproductive ability. This will be discussed in greater detail in upcoming chapters.

2.4   Neurological regulation of reproduction

We have discussed the basic mechanism of the reproductive cycle. Disruption of any part of this cycle can inhibit the normal function of the reproductive system. Particularly critical to this mechanism is the normal regulation and release of GnRH, the hypothalamic releasing factor responsible for the initiation and functional maintenance of the reproductive system in virtually all vertebrates. It is probably fair to say that GnRH is under greater regulatory control than any other molecular component of the reproductive system.

2.4.1 Gonadotropin-releasing hormone

Peptide hormones typically occur in families of related peptides as a result of the gene expansion mechanisms discussed earlier. As genes duplicate and change during evolution they give rise to families of related peptides and proteins. But GnRH is unusual among neuropeptides in that it does not appear to be related to any other known peptides outside of its immediate peptide family. Nevertheless, depending on the species of chordate, more than one form of GnRH may exist. For example, the terrestrial vertebrates appear only to have two forms of GnRH that have been highly conserved. Other chordates may have as many as four or five types of GnRH. For the sake of simplicity, we will focus on two main forms. One form, called GnRH-I, is found in the hypothalamus and telencephalon and is the primary releasing factor for the regulation of the gonadotrophins, particularly in terrestrial chordates. A second form, GnRH-II, is found in posterior or inferior parts of the brain, particularly in the midbrain, and has been implicated with the regulation of reproductive behaviour.

GnRH-I is also unique in that it appears to be the only hypothalamic releasing factor that migrates into the forebrain from outside of the brain during embryogenesis, specifically in the developing nasal and olfactory regions. Disruption of this migration leads to improper functioning or even complete inhibition of the reproductive cycles in the adult. Kallmann's syndrome, for example, is a condition in humans in which the embryonic olfactory tissue, the olfactory placode, fails to develop. Affected individuals have no sense of smell and abnormal reproductive cycles.

The migration of GnRH from the olfactory system to the brain is conserved in all vertebrates and may reflect an evolutionary connection between olfaction and reproduction. This relationship is important as the olfactory sense was probably one of the first sensory systems to evolve. Because reproduction is critical to the survival of a species, coupling of this sensory system with the reproductive system was necessary to coordinate reproductive physiology with events in the environment. In many species a sense of smell is particularly important to detect the presence of a mate, for example in those which use the perception of chemical signals (pheromones) released into the environment. In more neurologically complex organisms possessing multiple highly developed sensory systems, for example, primates, the role of the olfactory system becomes less dominant.

In mammals, the migration of the GnRH-I cells from the nasal placode to the front (rostral) part of the forebrain is a journey involving many chemical and hormonal signals to ensure that the cells reach their final destination. GnRH-I cells migrate across the nasal region on axons originating from cells in the olfactory pit and vomeronasal organ, which are organs associated with the detection of scents and pheromones. A number of chemical factors secreted by surrounding cells, for example, GABA, peripherin, nasal embryonic LHRH factor and N-CAM, play a role in the guidance, targeting and differentiation of the GnRH neurons during their migration.

Less is known about the function of GnRH-II. Despite the presence of GnRH-II in almost all species of vertebrate, establishing a clear function for this peptide has proved problematic. Previous studies have suggested it could act as a neuromodulator or neurohormone regulating reproduction and associated behaviours. Recent studies have implicated GnRH-II as a factor integrating mating behaviours and energy intake. For example, in female musk shrews, GnRH-II, but not GnRH-I, can enhance mating behaviour in animals that are food-restricted. Moreover, this effect appears to be mediated by the type II GnRH receptor. Studies of many chordates indicate that GnRH-II is primarily produced in a group of cells in the ventral midbrain (tegmentum), specifically in an area called the nucleus of medial longitudinal fasciculus. Yes, it is a bit of a mouthful, but it is important as we will discuss in the next chapter.

GnRH-II has an order of magnitude greater affinity for the type II receptor than for the type I receptor. This means that it binds much tighter to the type II receptor than to the type I receptor. The type II receptor is also expressed more widely throughout the brain than the type I receptor, being found preferentially in the amygdala, although expression also occurs in the hippocampus, substantia nigra, subthalamic nuclei and spinal cord. About 69% of LH-expressing cells of the sheep anterior pituitary express the type II receptor. In monkeys, GnRH-II has been found in hypothalamic and pituitary stalk regions, suggesting a potential role in the regulation of the pituitary gonadotrophins. In rams, GnRH-II releases a greater FSH to LH ratio than GnRH-I. Thus, GnRH-II may act in part as an FSH-releasing factor. Unlike the GnRH-I system, the GnRH-II neurons appear to arise locally around the third ventricle during embryonic development. From there, they undergo a short migration to the regions around the nucleus of the medial longitudinal fasciculus.

Despite the conservation of the GnRH-II structure throughout metazoan and particularly vertebrate phylogeny, its function as a separate and distinct signalling system is being lost in some lineages. The GnRH-II gene appears to have been silenced in mice, whereas in humans, chimpanzees, cattle, sheep and rat, the GnRH-II receptor has been silenced or disrupted. However, a number of other primates such as marmosets, African green monkeys and rhesus monkeys do possess functional GnRH-II receptors, suggesting that the silencing does not occur along phylogenetic lines. Thus, in general, mammals appear to be evolving to a single GnRH system, although it is not clear why this is happening.

2.4.2 Gonadotrophins and their regulation

The gonadotrophins, synthesized and released from the anterior lobe of the pituitary gland, are the peptides responsible for carrying out the systemic and gonadal actions associated with the reproductive cycles. In terrestrial vertebrates, the main pituitary gonadotrophins are FSH and LH. In ray-finned fishes, gonadotrophin I is orthologous to FSH, whereas gonadotrophin II is orthologous to LH. Although the piscine gonadotrophins have structural similarity to the corresponding tetrapod gonadotrophins, there are significant functional differences among fishes and for many species. For example, it is unclear how much gonadotrophin II participates in the regulation of the reproductive cycles. The gonadotroph cells of the pituitary express both gonadotrophins. Some of the secretory granules contain both gonadotrophins, whereas other secretory granules may contain either one gonadotrophin or the other. In mammals, the gonadotrophs are found scattered throughout the anterior pituitary gland. In teleost fishes, the gonadotrophins are generally clustered together in the anterior gland. This modification helps in fishes, because they receive direct input from the hypothalamic GnRH neurons.

GnRH is released into the portal system in a pulsatile manner. Low-frequency pulses favour the release of FSH whereas high-frequency pulses promote LH release. There are a number of neuroendocrine factors that act to modulate GnRH pulse frequency to ultimately affect gonadotrophin secretion. The pulsatile nature of GnRH release appears to be an intrinsic property of GnRH neurons. Coordination of secretion between GnRH neurons may be achieved by reciprocal axo-dendritic connections. Some modelling studies have suggested that coordination of release could be achieved with as few as 3% of cells making connections with one another. GnRH neurosecretory terminals converge at both the median eminence and organum vasculosum of the lateral terminalis (OVLT) and it is likely the regulation of coordination occurs in these regions. The pulse characteristics of GnRH are modified considerably by a number of other factors, particularly the gonadal steroids. However, a number of other neurohormones, including neuropeptide Y (NPY), noradrenaline, GABA, glutamate and nitric oxide have all been implicated in the regulation of GnRH pulsatility.

Although the simple model is that GnRH stimulates the release of the gonadotrophins, the mechanism varies considerably in different species of chordate. In rats, GnRH has been shown to increase α subunit, LH β and FSH β mRNAs, although the model for GnRH-mediated gonadotrophin release has been based on a single molecular variant of GnRH. In fishes, two and sometimes three variants of GnRH may play a role in gonadotrophin release. When gonadotrophin release is modulated by more than one form of GnRH, then the expression and release of the GnRH isoform that is used may be regulated in a seasonal manner, or by the stage in the reproductive cycle. In the frog, *Rana esculenta*, there are also two molecular forms of GnRH. The annual reproduction period in this species is characterized by long periods of GnRH accumulation followed by long periods of GnRH release. GnRH-II immuno-reactivity is high in the medial septum and low in the rostral region of the anterior preoptic area, whereas the opposite is true for GnRH-I. However, in amphibians, the expression of GnRH-I and -II tends to be found in overlapping regions of the brain. In some birds, GnRH-II may also contribute to the regulation of pituitary hormones, although its role varies with respect to season and breeding cycle.

2.4.3 GnRH regulation and onset of puberty

Puberty is the time when the reproductive system matures to an adult form. During puberty, the GnRH neurons are activated from a quiescent state to initiate the onset of the reproductive cycles. In mammals, GnRH neurons are active during the later stages of fetal development. In males, the GnRH neurons are turned on near parturition, activating the testes to produce the perinatal concentrations of testosterone required for the masculinization of the brain, whereas in females the GnRH neurons remain much quieter during fetal development, infancy and childhood.

After birth, the GnRH neurons are inhibited by neural factors and remain more or less quiescent until puberty, when they lose this inhibition and become active again. During the quiescent phase, GnRH suppression appears to be achieved, in part, by a direct inhibitory action of both NPY and GABA neurons on the GnRH neuron. In addition, NPY may also inhibit the stimulatory actions of glutamate-containing interneurons on the GnRH neurons. Puberty is initiated by the removal of these inhibitory mechanisms by both neural and endocrine factors. Increasing plasma leptin concentrations act to inhibit the direct NPY inhibition on the GnRH neuron and the indirect action via the glutamate-secreting interneurons. Additional neural factors appear to be responsible for inhibiting the GABAergic neurons, and stimulating the gluta-mate interneurons.

In the last ten years, a number of hormone systems have been added to our understanding of the regulation of puberty. Kisspeptin is a peptide hormone that regulates many of the positive and negative feedback signals relayed to the GnRH neurons and is also implicated in carrying signals associated with metabolic activity mediated by the hormone leptin, for example. Neurokinin B, another peptide hormone, colocalizes with kisspeptin in many cells in the brain that regulate the pulsatile release of GnRH from the median eminence. However, despite the importance of these hormones in the regulation of GnRH release, we have little understanding how stress affects the regulation of kisspeptin and neurokinin B. In fact, it may turn out that many of the actions of stress are mediated through these hormones. Although these mechanisms have been best studied in mammals, the basic mechanism is probably similar in all chordates but differs with respect to some of the details of the regulatory process.

2.5   Summary

Normal reproductive processes are regulated by a complex interplay of chemical factors released by the brain, pituitary gland and the various reproductive organs. In chordates, the primary hormones responsible for this effect are GnRH in the brain, the gonadotrophins in the pituitary and the sex steroids released from the gonads. The development of the reproductive system follows three distinct phases: the embryonic development of the germ cells and reproductive system, the neurological sexual differentiation of the brain, and the final maturation of the reproductive system during puberty. The normal secretion of GnRH-I is critical for the activity and maintenance of the reproductive system. GnRH-II is a second form of GnRH in most vertebrates and is also implicated in the regulation of reproduction and associated behaviours. Interference in the normal development of any of these processes can lead to disruption of the reproductive system.

3

The physiology of stress: why too much stress stops us from doing things we enjoy

Now all the soul's modifications do seem to involve the body – anger, meekness, fear, compassion and joy and love and hate. For along with these the body is to some degree affected. An indication of this is that sometimes violent and unmistakable occurrences arouse no excitement or alarm; while at other times one is moved by slight and trifling matters.

Aristotle, *De Anima* (fourth century BC)

3.1 Introduction

The stress response is frequently misunderstood. One of the questions that often arises in popular writings is, if the stress response is so good for you, why does it make us feel so miserable? Isn't something that is good for us supposed to make us feel good? There are a couple of ways we can consider this. Pain, for example, is uncomfortable. If it makes us feel good, we would do more of whatever was causing us pain. Sticking your hand in an open fire is damaging to your health. Pain makes us feel uncomfortable, so we don't do it again. The same thing occurs with the stress response. Pain can be thought of as one element of the stress response. There is a limit to the amount of stress we can handle. Similarly, there is a limit to the amount of pain we can tolerate. Before we experience pain, we might experience some pressure or temperature differences. Some of these feelings might even be pleasant initially. However, if the stimulus increases in intensity to the point that the stimulus produces damage to the organism, then pain ensues. Or, using a stress example, once the

Sex, Stress and Reproductive Success, First Edition. David A. Lovejoy and Dalia Barsyte.
© 2011 John Wiley & Sons, Ltd. Published 2011 by John Wiley & Sons, Ltd.

stress reaches a particular level – the point at which it threatens our survival – then we perceive the stress response as being uncomfortable.

The first elements of the stress response are physiological and, in general, the organism experiencing these is usually unaware of this phase. The stressors will be dealt with at the cell and tissue level. If the intensity increases, then the organismal response will become involved and additional neurological, endocrine and immunological responses become recruited. At this stage, humans experience 'anxiety'. Anxiety is our perception that there is a potential threat that requires our conscious attention, although we may not be consciously aware of what it is. Should this threat continue and the stress is allowed to persist despite our attempts to solve the problem, then behavioural responses become involved, and then a depression in behaviour occurs that essentially tells us, 'look if you don't stop what you are doing, we are going to take away all of the motivation from doing anything'. We then tend to feel tired and lose interest in the things we enjoy; we might stop eating, for example.

Although a strict physiological approach may be useful for many non-human organisms, humans tend to be a bit stubborn when it comes to stress. My father once complained that his doctor's advice did not reflect reality. My father was working at three jobs, had three young children and had just bought his first house. He was exhausted trying to make ends meet. 'Take a vacation' suggested his doctor. 'If I take a vacation, I will not be able to work', he replied, 'then I won't get paid and I'll be unable to meet my mortgage payments. Then my children will not have a place to live.' But we all continue to do such things. We need to work to get all of those things we think we need. We don't need most of these things, of course. We just think we do. The mismatch between our abilities and our real or perceived needs is what generates stress. But this is distinctly human. Animals tend to be much more practical.

3.2   Anxiety and the evolution of the stress response

There are physiologically useful reasons why such a system evolved. Consider an organism, say a small rodent. It is hungry and needs to get food. In order to do so it must leave the safety of its den. However, it can smell the fox lurking outside. It knows that the scent of the fox is a warning about death and so it remains in the safety of its burrow. The hunger will be reduced for the moment. If the fox remains and our little rodent becomes hungrier, then the two opposing forces begin to grow. The rodent is hungry but safe. In order to eat it must risk its life. Eventually the fear of the fox or the hunger will dominate. However, during this period of indecision between two different needs, the animal may experience something resembling anxiety.

If the animal was not hungry and was quite content to lounge around in its den, it probably would not experience the same level of anxiety. This brings up the concept of motivation as a determinant of action. A greater level of stressors may be endured as the motivation increases. However, anxiety may endure because of this mismatch between need and potential harm. This will be discussed in more detail in later chapters.

So, this feeling of anxiety is the perceptual manifestation of the activation of the 'fight or flight' syndrome we described in Chapter 1. Depending on the level of anxiety, an animal may experience the fight or flight response, or it may become paralysed with fear. During this acute fear, the animal may experience a slowing of heart rate, a lowering of blood pressure and weakening of muscular tone. Sometimes during an acute bout of intense fear, uncontrolled micturation (urine release) and defecation may occur. Then sweating, heart rate and blood pressure may increase, combined with a number of hormone changes, including increases in adrenaline, noradrenaline, cortisol, growth hormone and prolactin. In males, this may be accompanied by a decrease in testosterone.

We all have friends that we can categorize as high- or low-anxiety types. We might have one friend who panics when he loses his keys and another who spends her spare time skydiving or bungee-jumping. One friend might be dynamic and outgoing at parties whereas another avoids them. There is great variability in people's willingness to tolerate anxiety and, as you might expect, the level of occurrence of anxiety in a human population follows a normal or Gaussian distribution (bell-shaped curve) as it does in many other chordate species. The widespread occurrence of anxiety among vertebrates indicates that this trait has been selected during evolution because it confers a survival advantage in those species. In 1908, psychologists Robert M. Yerkes and John D. Dodson published their theory that performance is influenced by the level of arousal. This has become known as the Yerkes–Dodson Law, which states that a small amount of anxiety can improve performance whereas high anxiety can decrease performance. Imagine you have arrived home and are opening your front door with a set a keys – a simple task that does not require much thought. Now, imagine that as you approach your front door, an aggressive dog living across the street has escaped and is now running towards you, barking, with his fangs glistening in the sunlight. Suddenly, the action of opening your door becomes a bit more of a challenge; you fumble, mix up the keys and put the wrong key in the lock. That is a manifestation of the Yerkes–Dodson Law. Similarly, mild stress and anxiety can improve performance in some situations. In our scenario above, perhaps it is particularly cold outside and you are underdressed. So, in your dash from the car to the house, you have your keys ready and in a single motion insert the key and open the lock.

The Yerkes–Dodson Law also applies to sex. Work-related anxieties, parental pressure to produce lots of children, personal pressure to impress our partner can all affect our performance during sex. Just remember what it was like when you were a teenager.

Now why has anxiety endured in evolution? Psychologists have pointed out that anxiety alters the quality of how we gather, interpret and deal with information and events around us. We refer to these processes as cognition. For example, high-anxiety people tend to focus on the resolution of threat-associated stimuli, whereas low-anxiety individuals show a preference for the processing of non-threat-associated stimuli. Among university students, a high-anxiety individual may need to focus on finishing a term paper before going out with friends. A low-anxiety individual may opt to hang out with friends then deal with the essay. As we discussed in the first chapter, the stress response prepares the individual for fight or flight. In preparation of this it, vigilance and general awareness of the surroundings is increased. Ambiguous stimuli may be perceived as threatening by a high-anxiety individual. Thus in some cases, a higher anxiety individual may be more sensitive to the perception of danger than a low-anxiety individual. A more sensitive danger-warning system may allow this individual to escape from danger before some of their associates are aware of the danger. If this improves survival, then this trait can be selected for. However, there is a downside. If the level of arousal is too great, and the individual is overly sensitive to potential danger, then this can inhibit other physiological needs such as food and sex. Sexual performance with a partner would hardly be optimal if our partner jumped up at every small noise to check to see if anyone was coming to disturb them. Recalling my teenage courtship years, I can sympathize with this.

Anxiety has also been implicated in the promotion of social cohesiveness and emotional bonding, both between progeny and parents and between adults. Essentially, this comes down to a safety in numbers argument. If a predator is lurking and you are alone, the only prey, your chances of survival are much less than if you are standing around with a hundred friends. Thus, anxiety is our perception of the warning system that we are in potential danger. However, too much anxiety, while increasing our perception of danger, could interfere with our interactions with others, with digestion and reproductive processes and ultimately reduce our ability to reproduce. Too little anxiety may lead to a reduced perception of danger, leading to personal injury or death and therefore also a reduced capacity to produce progeny. In our modern society, in anxiety-avoiding people this might manifest as social isolation, lost opportunities, loss of motivation and lack of competition for jobs. Other people, in contrast, may engage in events or situations that increase anxiety, resulting in greater motivation to complete a task.

Humans have recognized anxiety early in our socialization. The word 'anxiety' itself comes from the early Indo-European word *angh*, which means to strangle or press tight, and also a burden and trouble. The concept of anxiety in various forms was described by Democritus, Plato and Aristotle. In the ancient texts *The Epic of Gilgamesh* from Sumer and Homer's *Iliad* from ancient Greece the protagonists express their anxieties. Many of our writings, performances, expression of art and music have been an attempt to resolve our anxiety. Anxiety can be stimulating for us. Where would our interest lie in the story of *Romeo and Juliet*, if the Montagues and Capulets simply said, 'Right. Enough fighting. Let's all get along. You two youngsters should get married and have children as soon as possible.' These artistic expressions of emotion are a socially acceptable way to regulate and rationalize the anxiety we may experience in the wake of a set of stressful events. However, anxiety becomes abnormal when its intensity and duration are not proportional to the level of the perceived threat. Emotions associated with anxiety can include worry, fear, feelings of doom and anger. Depression may occur when a person loses their ability or confidence to control anxiety.

Do non-human animals experience anxiety? We have to remember that the concept of anxiety was developed by humans for humans to describe sensations that humans felt. On the basis of our understanding of human anxiety, we have tried to understand anxiety in other species. Because of this, and the emotions that are instrinsically tied up in our human experience of anxiety, our perception is uniquely ours and, therefore, our understanding of anxiety in other species is always going to be biased toward what we think anxiety is. However, we have a number of ways of examining anxiety. Certain behaviours are associated with anxiety. Avoidance of other individuals or the reaction in a novel situation can be a useful behaviour to predict the presence of anxiety. The presence or absence of these behaviours can reflect the level of anxiety an animal is experiencing.

There are a number of neurological and endocrine responses associated with anxiety. In animals that are physiologically similar to us, similar responses can indicate whether anxiety is present. Countless studies performed over the last hundred years or so have provided substantial evidence that most, if not all, sufficiently complex vertebrates are capable of anxiety. Of course, we will probably never know how an individual animal perceives anxiety, but we can be reasonably confident whether a stress response is activated or not. Based on studies of neurological complexity there is evidence that the greater the complexity of the nervous system the greater the experience of anxiety. In a basic sense, this is logical. The more ways there are to perceive the world and remember the bad experiences the more it opens the door for numerous types of anxiety. We assume also that the simpler the nervous system, the lower the levels of anxiety an organism feels because there are fewer responses possible following a stimulus. In other words, I will tolerate a chipmunk pestering me for food at a campsite, but will not think twice about swatting a mosquito.

In this modern world, as our opportunities to experience wilderness are in decline it comes more important than ever to understand anxiety. Our perception of stress and anxiety is critical to understanding the impact we have on other species and their habitats and how we can manage our impact.

3.3   Stress, anxiety and the nervous system

In the last chapter we introduced elements of the brain and nervous system that are implicated in the regulation of reproduction. The situation with the stress response is more complex. Where the neurological components of the reproductive system are designed to focus primarily on the proper function of reproduction, the neurological components of the stress response system oversee all physiological actions, including reproduction, growth, digestion and cardiovascular activity. Moreover, the organism must integrate all of this information. At some point in this integration, the data will be forwarded to the reproductive system. The organism must be in as close to an optimal state as possible for reproduction to occur.

In humans, our development of treatments for many stress- and anxiety-related conditions has lagged behind those for most other types of diseases and illnesses. This is partly because there are often multiple regions of the brain playing a role in such disorders. To complicate matters further, two people apparently suffering from the same anxiety-related condition may have a different set of brain regions that are involved in the development of the illness. Now, before we all begin to panic with 'treatment of anxiety' anxiety, there are a number of regions of the brain that are common to most aspects of the stress response. Central to stress response physiology in vertebrates is the hypothalamic–pituitary–adrenal (HPA) axis (Figure 3.1). In fishes, this is known as the hypothalamic–pituitary–inter-renal (HPI) axis to reflect the different morphology of the adrenal gland in these species. Conceptually, this is similar to the hypothalamic–pituitary–gonadal (HPG) axis we described in Chapter 2.

After the appropriate stimulus, the hypothalamic peptide corticotrophin-releasing factor (CRF) is released into the vascular system of the median eminence (in most vertebrates) or by direct neurosecretory interaction (in most teleosts). This induces the pituitary corticotroph cells to release adrenocorticotrophic hormone (ACTH) into the systemic blood. Although this is the scenario in most species, there are some variations. For example, in most vertebrates, the vasopressin-like and oxytocin-like peptides also stimulate ACTH release. In some mammals, depending upon the type and duration of

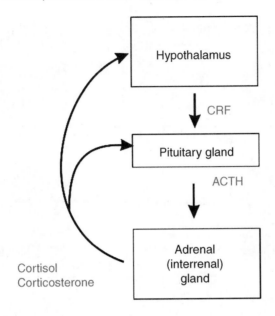

Figure 3.1 Hypothalamic–pituitary–adrenal/inter-renal axis

the stressor and the species, vasopressin may act as the primary ACTH-releasing hypothalamic peptide. In the rainbow trout, CRF and arginine vasotocin synergize to release ACTH, a situation similar to that seen in mammals.

The release of ACTH into the systemic circulation stimulates the synthesis and release of the glucocorticoids from the adrenal/inter-renal gland. This mechanism is well conserved in chordates. Although little is known about the HPA/I (hypothalamic–pituitary–adrenal/inter-renal) axis in cartilaginous fish, it is likely to be more or less intact because the essential components are present in more primitive fishes such as lampreys. Like oestradiol, progesterone and testosterone, the glucocorticoids are steroid hormones. The metabolic actions of the glucocorticoids are to provide the animal with enough energy to survive a stressful situation. Glucocorticoids possess a number of actions associated with the utilization and flow of energy. For example, the de novo synthesis of glucose, gluconeogenesis, is increased in the liver following mobilization of proteins from skeletal muscle and consequent deamination of amino acids that are released from protein breakdown. However, glycogen can also be deposited in the liver because of a glucocorticoid-stimulated increase in glycogen synthase reaction. In tandem with this, the breakdown of glycogen, the storage form of glucose is inhibited (Figure 3.2).

The actions of the glucocorticoids are based on a short-term need for energy. During a stress reaction, energy can be obtained via protein breakdown.

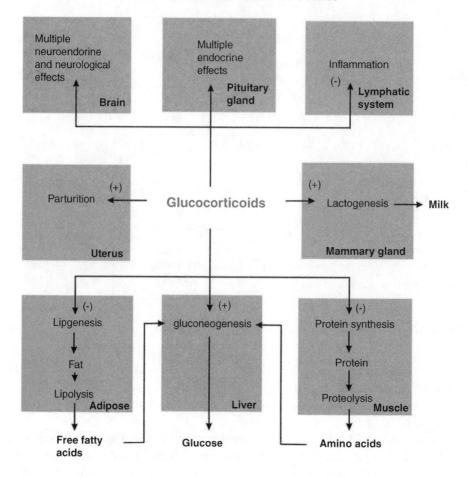

Figure 3.2 Actions of glucocorticoid hormones

Glucose stores in the form of glycogen in the liver are therefore protected. In birds, increased adrenocortical secretion in response to stress can redirect activity by facilitating an increase in food searching and food intake needed to meet periods of increased energy demand. In the rainbow trout, apoptosis and cell proliferation are also influenced by glucocorticoids. Low levels stimulate apoptosis whereas high levels stimulate proliferation.

3.4 Autonomic nervous system

The HPA axis is primarily a neuroendocrine mechanism, but during the activation of the stress response it acts in tandem with a component of the nervous system called the autonomic nervous system. The autonomic nervous system is an involuntary nervous system; the organism does not have direct conscious control over its activation. It consists of two main branches: the parasympathetic nervous system and the sympathetic nervous system (Figure 3.3). The parasympathetic nervous system acts to conserve and store energy and is associated with digestion and growth. The sympathetic nervous system, on the other hand, is involved in the expenditure of energy and responses to threat and homeostatic challenges.

The sympathetic nervous system, therefore, is the branch of the autonomic nervous system that is essentially responsible for arousal in vertebrates. Many of its effects are mediated by the catecholamine hormones adrenaline and noradrenaline. The synthesis of adrenaline begins with the amino acid tyrosine. This is taken up by the chromaffin cells of the adrenal medulla or the head kidney in fishes and converted to dihydroxyphenylalanine (DOPA), which is subsequently converted to dopamine. Through enzymatic action it is then converted to noradrenaline and then finally adrenaline. Secretion of adrenaline from the chromaffin cells is stimulated directly by acetylcholine release from preganglionic sympathetic fibres innervating the adrenal medulla.

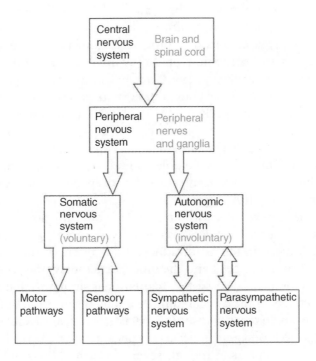

Figure 3.3 Main divisions of the nervous system. Modified after Brown (1994)

Thus the actions of the sympathetic and parasympathetic nervous systems are functionally antagonistic to each other. The sympathetic nervous system acts to increase the mobilization of energy and redistribute the blood supply – actions that are essential during a stress response. The parasympathetic nervous system promotes the digestion of food and storage of energy.

The sympathetic nervous system has two distinct branches. The sympathetic adrenal medullary branch is associated with the synthesis and release of catecholamines from the chromaffin cells of the adrenal medulla in vertebrates, or the head kidney of fishes. Adrenaline is the primary catecholamine released. The neural branch consists of the sympathetic nerve terminals impinging primarily upon the cardiovascular system. Noradrenaline is mostly released from these nerve terminals. In response to stress the sympathetic nervous system and adrenocortical (HPA) system are activated simultaneously, but the time-courses of their actions follow different routes. In humans, for example, the catecholamines released from the adrenal medulla and sympathetic nerves act almost immediately, whereas the adrenal glucocorticoids take effect about 20–30 minutes later.

If this is confusing, let's consider this on a practical level. Imagine that you have been studying for a final exam. You spent the last week preparing and now you are confident that you will ace it. You arrive at the examination room a few minutes before 1 p.m. when the exam is scheduled. However, the room is empty and there is a notice on the door indicating that the exam was at 9 a.m. You missed it! You feel an uncomfortable sensation in your stomach and begin to shake. This is the activation of the sympathetic nervous system. You suddenly see your career ahead of you vanish. Eventually, you gain composure and seek out the professor for your course. You explain your predicament to him. He listens impassively, then tells you he will think about it and let you know next week what his decision will be. During the next week, you begin to wonder if you will be able to graduate and anxiety begins to build up. If you cannot graduate, you will not be able to find work or go onto professional school. You find it hard to concentrate, eat or sleep. Now you are experiencing the activation of the HPA system.

Fish may not have to deal with final exams, but they do have to deal with many other challenges. The teleost version of the adrenal gland that produces adrenaline is located in the head kidney and consists of the catecholamine-synthesizing chromaffin cells and the glucocorticoid-synthesizing inter-renal cells. The autonomic nervous system regulation is similar to that in mammals. The primary effects of the catecholamines are on respiration, branchial blood flow, heart rate and oxygen transport capacity, as well as metabolic actions on free fatty acid and blood glucose levels via glycogenolysis in the liver mediated by β-adrenergic receptors, although several teleost species may also utilize α-adrenergic receptors.

Although we have focused on the hormones directly associated with the sympathetic nervous and adrenocortical systems, other hormones in the brain act to link these two systems and coordinate their responses in the face of stressful challenges. One such hormone is 5-hydroxytryptamine (5HT), also known as serotonin. 5HT is found throughout the brain and links the processing of many neurological systems. In humans and mammals it is associated with emotions and learning, among other things. Some evidence suggests that emotional stress is related more to the activation of the adrenal medullary branch of the sympathetic nervous system and a physical workload more to the activation of the neural branch of the sympathetic nervous system. For example, administration of 5HT agonists to the para-ventricular nucleus (PVN) of resting rats increases both the activity of the HPA axis and the adrenal medullary branch of the sympathetic nervous system, whereas administration of 5HT agonists to the PVN of exercising rats leads to a reduction of the activity of the neural branch of the sympathetic nervous system.

Emotional stress is accompanied by increased 5HT activity in the hypothal-amus. If a workload is of moderate intensity, then about 70% of the total energy expenditure is a result of free fatty acid breakdown. Lipolysis and free fatty acid production are maintained during long periods of stress by continued activation of the nervous branch. But for emotional stress, the increased CNS activity is entirely glucose-dependent, thus the adrenal medullary branch is preferably activated. It must be borne in mind, however, that the organization and function of these systems in mammals may differ considerably from those in other vertebrates.

3.4.1 Hormones related to corticotrophin-releasing factor

In Chapter 1 we discussed the concept of genome expansion and in Chapter 2 we mentioned that there were two different forms of GnRH. In contrast, CRF is part of a family four related peptides. It is acknowledged as the primary hypothalamic factor mediating stress-induced adrenocorticotrophin (ACTH) secretion from the anterior pituitary in vertebrates, but the majority of these investigations have utilized mammals. The CRF neurons that regulate ACTH release are found in the parvocellular cells of the PVN of the hypothalamus. These cells are sensitive to glucocorticoid feedback from the adrenal cortex and, therefore, form the neuroendocrine negative feedback loop resulting from high plasma concentrations of the glucocorticoids in response to ACTH stimulation. However, CRF is also present in a number of locations in the brain including the bed nucleus of the stria terminalis, central nucleus of the amygdala and in a number of other hypothalamic and brainstem regions.

Urotensin-I is a CRF-like peptide that was originally isolated from the urophysis of fish. Since its first discovery an orthologue, termed urocortin, has been cloned from mammal brain. Urotensin-I/urocortin peptides and CRF were formed as the result of a gene duplication. Although in fishes most of the urotensin-I in the nervous system is present in the urophysis, the majority of cells expressing the hormone in the brain occur within the nucleus of the medial longitudinal fasciculus of the midbrain (tegmentum) region. In mammals, urocortin is found in the Edinger–Westphal (EW) nucleus, a related region of the brain. The functional role of the midbrain urocortin/urotensin-I is not clear, but it does play a role in a number of stress and anxiety-related functions, including suppression of appetite. Peripheral urotensin-I and urocortin have both been implicated in cardiovascular responses.

Urotensin-I/urocortin and CRF may have complementary actions in different parts of the brain. Interestingly, CRF is found in regions of the brain that express GnRH-I and studies indicate that CRF may directly connect with the GnRH-I neurons to inhibit or modify the GnRH-I release. Urocortin and urotensin-I are also found near the cells that release GnRH-II and may play a role similar to that played by CRF, providing a direct regulation of GnRH-II. Thus at neural level and independent of HPA axis activation, both CRF and urocortin may inhibit some reproductive processes directly. As we pointed out in the previous chapter, the two GnRHs are involved in different aspects of reproduction and, therefore, the types of stressors that regulate them differ. The relationship they have with urotensin-I/urocortin and CRF may reflect these different stress pathways.

For the most part of this book we will focus on CRF as the central stress-associated neuroendocrine system. However, recent findings have indicated that the CRF system is more complex than we previously understood. The CRF/urotensin-I lineage is paralogous to a second lineage of CRF-like peptides. This lineage includes urocortin-2 and urocortin-3. It is not really clear how much influence urocortins 2 and 3 have on the stress response. At this point their putative role is based primarily on the structural similarity these peptides have with CRF and experiments that have been performed to show similar modes of action to CRF.

There are two types of CRF receptors. The two primary subtypes (R1, R2) show the greatest sequence similarity to the class II group of G protein-coupled receptors (GPCR). At least four isoforms of the CRF-R1 receptor have been identified and at least four subtypes of the CRF-R2 receptor. Unlike the R1 receptor, the R2 subtypes show considerable selectivity among the CRF-related ligands. The R1 receptors have been implicated in the onset of anxiogenic-type behaviours, whereas the R2 receptors have been associated with a number of physiological mechanisms outside the brain as well as neural activity. This receptor may be particularly important in the mediation of cardiovascular activity by CRF-related peptides.

3.5 Complementary physiological systems

Wouldn't it be nice if each physiological system in the body was responsible for one function? It would be so much easier to understand how the body works. However, the evolution of physiology systems is rarely so focused. Without getting into a philosophical discussion on the direction of the evolutionary process, we can state that any sufficiently flexible physiological system may be adapted over the evolution of a taxon to take on a variety of functions. We mentioned in Chapter 1 that the interaction of stress and reproduction can involve a number of other systems, including energy production, sensory perception and defence mechanisms, for example. Therefore, it is useful to understand some of the other systems that play a role in the stress response. Each of these systems will be associated with a different complement of the system but will also, as expected, vary considerably among species.

3.5.1 Caudal neurosecretory system

The activation and interaction of the sympathetic nervous and adrenocortical systems comprise the core physiological mechanisms associated with the stress response. In fishes there is an additional neuroendocrine system found in the tail called the caudal neurosecretory system (CNSS), which plays a role in certain types of stress. This system has been lost in the terrestrial vertebrates. As you will recall from Chapter 1, the nature of the stressor that an organism perceives is dependent, to a certain extent, on the environment in which the animal lives. Aquatic environments have a set of stressors that is distinctly different from that in terrestrial environments.

The CNSS or urophysis in teleosts (bony fish) has been implicated in the regulation of glucocorticoids. For example, removal of the urophysis in goldfish has been shown to increase plasma cortisol, possibly because of compensatory secretion of urotensin-I from the goldfish hypothalamus. This observation is supported by urophysis removal in a suckerfish, which results in elevated urotensin-I mRNA levels in the lateral tuberal nucleus (LTN) of the hypothalamus, while CRF mRNA in the preoptic nucleus and LTN remain more or less constant. However, urotensin-I may also stimulate the HPI axis directly at the level of the inter-renal tissue. Urotensin-I appears to stimulate cortisol secretion directly and to interact synergistically with ACTH to promote cortisol secretion in inter-renal preparations isolated from saltwater-adapted trout. In trout and flounder, urotensin-I injection *in vivo* leads to a dose-dependent increase in plasma cortisol secretion, suggesting that such responses are of physiological relevance.

Thus, the CNSS may provide a pituitary-independent route to the regulation of cortisol secretion, affording alternate, perhaps stimulant-specific mechanisms to modulate plasma cortisol. This may be particularly important in fish groups that lack the hypothalamic pituitary portal blood link and thus lack the complex regulatory input of a cocktail of hypothalamic factors delivered to ACTH cells in tetrapods. The fixed and direct innervation of ACTH cells of the fish pars distalis inevitably restricts the variety of stimulants or moderating factors that can affect ACTH, and thus inter-renal secretion. The CNSS provides an additional mechanism to allow other stimulant-specific inputs to modulate cortisol secretion.

3.5.2   Hypothalamus-intermediate lobe regulation

Melanocyte-stimulating hormone (MSH) is the principal hormone secreted from the intermediate lobe of the pituitary. In many non-avian and non-mammalian vertebrates, MSH is responsible for the colour change associated with background colour adaptation and camouflage. Melanin is a dark pigment found in some skin cells that, under the appropriate stimulus, can be dispersed throughout the cells as an adaptive response. For example, in humans, melanin dispersion is stimulated in response to UV light as a protective mechanism against these harmful rays. In fishes and amphibians, it may be released to darken the skin to allow the animal to be less visible against a dark background, such as occurs when the animal hides under rocks or vegetation.

This colour-change mechanism is associated with elements of the stress response. The MSH-secreting cells (melanotropes) are under the regulatory control of a host of stimulatory and inhibitory factors from the hypothalamus. The regulation of MSH from the intermediate lobe has been particularly well studied in frogs. In frogs and toads, a subset of NPY cells of the suprachiasmatic nucleus (SCN) in the hypothalamus directly innervate the intermediate lobe. In these cells, NPY is co-localized with dopamine and GABA and acts as an α-MSH release inhibiting factor. The melanotrope inhibitory neurons of the SCN appear to receive direct photic input from the retina. Thus, this pathway appears to be associated, in part, with the perception of background colour.

A number of physiological systems in vertebrates employ CRF and NPY in opposition to each other. A similar situation appears to occur with respect to MSH-release regulation in the intermediate lobe. Thus, in addition to CRF- and urotensin-I-like peptides stimulating the HPI axis, these peptides also stimulate teleost pituitary melanotropes that release numerous peptides including ACTH and other hormones. Although not part of the intermediate lobe, the melanin-concentrating hormone (MCH) plays a complementary role to MSH with respect to pigmentation. MCH reduces pigmentation in the teleost integument but also plays a separate role in the regulation of the stress response. In both rats and trout, exogenous MCH decreases ACTH release as well as inhibits melanin dispersion in the melanophores of the integument. In mammals, MCH plays a less central role in the regulation of colouration, and has been implicated as part of the stress-response system regulating appetite and sleep. It has also been recently found to regulate aspects of learning and memory.

3.5.3 Thyroid hormones

The thyroid hormones are also important during the activation of the stress response. The central neuroendocrine circuit associated with thyroid hormone regulation is similar to the HPG and HPA axes. In the hypothalamus–pituitary–thyroid gland (HPT) axis, control begins with the release of thyrotrophin-releasing hormone (TRH) from the PVN of the hypothalamus. Like GnRH and CRF, it is released into the pituitary portal system in terrestrial vertebrates, where it travels to the anterior lobe to cause the release of thyroid-stimulating hormone (TSH) or thyrotrophin. TSH is released into the bloodstream to cause the synthesis and release of the thyroid gland hormones, mostly, T3 (triiodothyronine) and T4 (thyroxine). These hormones, although not steroids, act like steroids and bind to receptors that are similar to steroid receptors.

Thyroid hormones are important for the regulation of metabolism and oxygen usage in cells. Thus, when increased activity of cells and tissues is required, thyroid hormones are secreted. Because their receptors are transcription factors, thyroid hormones can regulate genes directly. They are also important during development and are implicated in the differentiation of cells into new cell types.

During the stress response, the thyroid axis is stimulated in order to meet the extra metabolic demands required during the 'fight or flight' response and may remain elevated while the HPG axis is active.

In some species, such as frogs and perhaps fish, CRF is tied to the HPT axis by direct stimulation of TSH from the pituitary gland. In mammals, however, CRF appears to stimulate the HPT axis by first stimulating the release of TRH from the hypothalamus. Centrally and peripherally injected CRF induces a dose-dependent hypothermia in rats, although this mechanism may also involve dopamine and noradrenaline. The effects of CRF peptides on thermogenesis may be related, in part, to their interaction with TRH. In mammals, CRF and TRH are closely situated within the PVN and appear to regulate each others' release of reciprocal connections, thus some of these thermogenic responses may be mediated by TRH and the thyroid axis.

During periods of HPA activation, the HPT axis is also frequently activated. Typically, circulating T3 levels are highly regulated by peripheral tissues, however stresses such as trauma, burns, infections, diseases, myocardial infarction, chronic diseases or metabolic disorders will elicit a thyroid hormone response resulting in, paradoxically, low serum T3 levels. This is because of the decreased conversion of T4 into T3. T4 is instead directed toward alternative metabolic pathways. The resulting lowered T3 levels allow for conservation of protein and also the initiation of gluconeogenesis.

There are three main potential mediators of low T3/T4 levels. First, infection or trauma lead to increased cytokines that induce low levels of thyroid hormones. Secondly, surgery or burns promote an increase in serum cortisol concentrations that cause low T3 levels. Thirdly, nutritional status is another factor capable of inducing serum thyroid hormone changes. Either fasting or uncontrolled diabetes mellitus can initiate a fall in serum T3 levels.

3.5.4 Arginine vasopressin as an ACTH-releasing factor

In most mammals there is a subpopulation of cells in the PVN that secretes both CRF and vasopressin into the median eminence. CRF actions may be potentiated by the presence of vasopressin. In response to the appropriate stressor vasopressin may also be released from magnocellular cells of the supraoptic nucleus. Thus, there are distinct subpopulations of vasopressin-secreting cells that are secreted into the portal system of the median eminence, and also that directly project to the neurohypophysis where vasopressin is stored. Neurohypophysial arginine vasopressin is secreted directly into the systemic blood and plays a role in osmoregulation. In fishes, CRF appears to act as the primary releasing factor but, like mammals, the vasopressin (vasotocin)-related peptides also act as ACTH-releasing factors under some conditions. Arginine vasotocin and isotocin can produce additive actions but do not appear to potentiate the ACTH response to urotensin-I or CRF in goldfish, indicating that their physiological roles likely vary among species.

3.5.5 Proopiomelanocortin and ACTH

The structure of the ACTH precursor proopiomelanocortin (POMC) along with genes that are structurally related is another example of gene expansion events in evolution that have increased the complexity of the possible responses to different types of stressful stimuli. POMC is the precursor not only for ACTH, but also for MSH (melanotrophin) and β-lipotrophin. There are several variants of these peptides among the different vertebrate species. Five subtypes of receptors have been isolated that bind MSH- and ACTH-related peptides, named MC1–MC5. The MC1 receptor has been implicated in pigmentation and may also mediate the anti-inflammatory actions of the melanocortins. The MC2 receptor is found exclusively in the adrenal glands and mediates the glucocorticoid-synthesizing and releasing effects of ACTH. The MC3, 4 and 5 receptors are primarily found in the brain, although MC5 is found in a number of tissues as well. The MC3 and MC4 receptors have been implicated in the regulation of feeding, appetite and metabolic rate.

POMC belongs to a family of four related prohormones. Proenkephalin is the precursor for the opioids leu-enkephalin and met-enkephalin, prodynorphin is cleaved to yield a number of dynorphin-related peptides, and pro-orphanin is the prohormone for orphanin FQ. All peptides act upon the opioid receptors μ, κ and δ, with the exception of orphanin FQ that acts on its own specific receptor to mediate hyperalgesia. These hormones have been implicated in a variety of functions including algesia, thermogenesis, feeding, learning and memory, sexual function and behaviour, cardiovascular and respiratory actions.

3.6   Integration of HPA/I components with other systems

At this point, you might be feeling overwhelmed by the sheer number of hormones and their receptors that are associated with the stress response. Remember that there are numerous sensory stimuli that can indicate a threat to the organism. Depending on the stimulus, each will have their own complement of chemical messengers to inform the rest of the body that it is under threat. Although we have described these systems as if they have their own independent function as the result of a particular threatening stimulus, their actions are highly integrated with other hormone responses. We tend to describe the function of hormones in this way because we only study a few responses at a time. The reality is very different and several physiological systems may be activated as a result of a single stimulus. Thus, a single integrated response to a stimulus may include hundreds or thousands of individual reactions that affect numerous physiological systems.

Although the stress response is highly integrated with other physiological systems, in general, such a response acts to inhibit parasympathetic-related functions and stimulate sympathetic-related functions. When organisms are confronted with an event that threatens their survival, the immediate goal is not to stop and eat or drink, but rather deal with the threat. If you survive a threat, then you can always eat later. But what is important is that the water, ions and nutrients you already possess in your body are used most efficiently. The hormones of the HPA axis play a number of roles in the regulation of feeding and drinking. However, how the HPA axis is integrated with other physiological systems will vary considerably among different species. The type of integration will ultimately reflect the habitat and niche that the animal has become adapted to.

Diuresis, or the regulation of water ingestion (i.e. drinking), is under a complex set of regulatory controls that are essential to both the regulation of stress and reproduction. Water loss is a much more profound stressor than fasting and nutrient loss. Water, like nutrients, must be constantly replenished. Throughout the day, water is lost through respiration, micturation, defecation, salivation and sweating, although the relative amount of loss varies among species. During the various stages of reproduction, extra water is required for the development of tissues associated with pregnancy such as mammary glands, the uterine lining and placenta, and to a lesser degree, gamete maturation. Lactation is particularly sensitive to water loss.

Water loss may occur as a result of reduced availability, heat, exercise or increased metabolic activity and tissue growth. It may also be promoted as a response to various stressors, by excess muscle or neurological activity leading to increased metabolism and, hence, heat production. In mammals, water loss in the form of increased sweat or salivation occurs as a direct response to heat.

Other species may have to contend with other stressors that induce water loss. In marine fishes, for example, the main cause of water loss is the high levels of salinity in the surrounding water, which promotes loss of water through the skin and gills.

In general it is difficult to separate the physiological need for water ingestion from feeding and the need for nutrient ions. Because there needs to be a balance between the amount of water and ions in tissues, both aspects tend to be under a set of complementary regulators. Ions such as Na^+, Cl^-, K^+ and Ca^{2+} are particularly important for the normal function of cells and tissues. These are obtained mostly through the diet, although marine fishes may gain some through the ingestion of seawater.

The effect of HPA-associated hormones on Na^+ ingestion varies among species. For example, glucocorticoids by themselves do not induce Na^+ uptake in rats, but do when they are administered to rabbits or sheep. Rats subjected to immobilization stress, however, will ingest more Na^+ when released than would be expected from compensating for their sodium loss, indicating that a stressor can induce Na^+ ingestion in this species. Although stress-induced Na^+ ingestion may be glucocorticoid-independent in rats, evidence suggests that CRF itself can elicit Na^+ ingestion that is independent of Na^+ loss. Similarly, if Na^+ is increased by hypertonic saline infusion in rats, CRF is decreased in the central nucleus of the amygdala and in the PVN. Such a finding suggests that CRF may regulate the hormone angiotensin II with respect to its role in Na^+ ingestion. The central nucleus of the amygdala plays a key role in the integration of signals associated with stress, and osmoregulatory processes. Thus, HPA hormones may not necessarily have a direct effect on Na^+ ingestion or osmoregulatory processes, in general, but rather modulate the actions of other hormone systems, such as angiotensin II and aldosterone, which are directly involved.

Angiotensin II (a peptide) and aldosterone (a steroid) are essential hormones involved in the regulation of water and ion intake and excretion. Together, subthreshold doses of angiotensin II and a mineralocorticoid (deoxycorticosterone) can elicit Na^+ ingestion, indicating that the two systems may be synergistic. Glucocorticoids and mineralocorticoids are synergistic in rabbits at stimulating sodium ingestion. One mechanism for this synergism may occur at the receptor level. For example, mineralocorticoids increase the number of angiotensin receptors in the brain. This is mediated by the type I receptor. Cortisol potentiates angiotensin II-induced water intake as well as mineralocorticoid-induced Na^+ intake via the AT2 receptors. It may also increase the amount of angiotensinogen synthesis (found in glial cells) by binding to type II glucocorticoid receptors. Cortisol binding to the type II receptors can also increase type I glucocorticoid receptor expression.

Food deprivation in rats will activate the HPA axis and, reciprocally, activation of the HPA axis attenuates feeding behaviour. Low concentrations of corticosterone will stimulate food intake, high concentrations will inhibit food intake. The hormones of the HPA axis can regulate a number of regions in the brain that affect eating. For example, lesions to the ventromedial hypothalamus can induce rats to overeat and become overweight. Removal of the adrenal glands, which removes the source of glucocorticoids, abolishes the tendency to overeat whereas cortisol replacement reinstates it. The glucocorticoid-mediated actions on feeding appear to be mediated primarily via the type II glucocorticoid receptor. The rise in plasma glucocorticoids may impart preferences in the diet. For example, protein intake is unaffected by corticosterone treatment although fat intake is. Short-term intake is associated with the type II cortisol receptors linked to carbohydrate intake. Type I receptors may also play a role in fat ingestion and long-term regulation of energy balance.

The actions of CRF and NPY are antagonistic in appetite regulation. CRF inhibits appetite and feeding, whereas NPY tends to promote feeding. Adrenalectomy can abolish NPY-induced feeding and cortisol replacement reinstates it. Stimulation of the type II glucocorticoid receptors using specific agonists increases NPY gene expression in the basomedial hypothalamic region or arcuate nucleus. Both regions send projections to the PVN. NPY is regulated under conditions in which food access is compromised and rats are hungry. Elevated NPY may also be associated with the enjoyment of rewards such as food intake.

Insulin and leptin are both part of a feedback loop linking feeding and the HPA axis. Insulin is a hormone produced in the pancreas that promotes the transport of glucose into tissues. Leptin is a hormone produced by adipose (fat) tissue that informs other tissues about the status of adipose storage. Thus both hormones play an important role to inform the organism about the status of energy and food intake. Adrenalectomy will reduce insulin secretion but cortisol will restore or increase it. Insulin receptors are present in PVN and arcuate nucleus, where they modulate the sensitivity of leptin and other factors on CRF and NPY expression, respectively. Leptin has modulatory actions on the HPA axis. It alters the expression of CRF in the hypothalamus, interacts with ACTH at the adrenal level and is regulated by glucocorticoids.

Plasma leptin and cortisol concentrations show inverse circadian rhythms to each other and may receive photoperiodic input independently from each other. For example, glucocorticoids play a modulatory but not essential role in generating the leptin diurnal rhythm. In humans glucocorticoids can act directly on adipose tissue and increase leptin synthesis and secretion. They appear to act as one of the key modulators of body weight and food intake, promoting leptin secretion by adipocytes, limiting central leptin-induced effects and favouring those of the NPY system.

3.7   Prolactin and stress

Prolactin is an interesting hormone in that it is involved in both the stress response and reproductive processes. It is probably best known as a hormone that regulates lactation in mammals but is also important as a growth factor and for regulating water and ion balance in tissues. Prolactin may play an important role in the activation of the HPA axis. Restraint stress can decrease levels of plasma prolactin in rats and sheep. However, although in rats injections of CRF stimulate the release of both ACTH and prolactin from the anterior pituitary, its presence may not regulate pituitary prolactin mRNA synthesis. Furthermore, in humans and sheep, CRF does not have a direct effect on pituitary prolactin, but prolactin may instead regulate the release of CRF via a centrally located mechanism.

Prolactin has number of modulatory actions on the immune system. Exposure to short days reduces blood prolactin levels in mammals and has a pronounced effect on immune function in a variety of species. In general, prolactin enhances normal immunological activities, but can also compromise immune function, particularly at high or low circulating levels. Since exposure to long days increases prolactin levels it could be the source of some of the seasonal changes that have been shown to occur in immune function in mammals. Removal of the pituitary gland of rats results in compromised humoral and cell-mediated immunity, which is restored by prolactin-replacement therapy. The immunological effects of prolactin might also interact with the immunological effects of steroid hormones to mediate seasonal fluctuations in immune function. In lymphocytes, prolactin increases hormonal and cellular immunity, reversing anaemia, leukopenia and thrombocytopenia caused by removal of the hypophysis. Prolactin also increases antibody formation. In addition to stimulating cell proliferation, prolactin inhibits apoptosis of lymphocytes. In the mammary gland, prolactin enhances immunoglobulin A (IgA)-secreting plasma cell activity.

3.8 Summary

The physiological mechanisms of the stress response have evolved for organisms to anticipate and prepare themselves for challenges that may threaten their survival. When a stressor is perceived, but the ability to resolve the stress is denied, we experience anxiety. The hypothalamus–pituitary–adrenal axis and the sympathetic nervous system are the two neuroendocrine systems responsible for mediating the stress response. Corticotrophin-releasing factor (CRF), adrenocorticotrophic hormone (ACTH) and the glucocorticoids cortisol and corticosterone are the main hormones associated with the HPA axis, whereas adrenaline and noradrenaline are the main hormones associated with the sympathetic nervous system. The HPA axis and sympathetic nervous system closely interact with the thyroid hormone system. The goal of all systems is to provide sufficient energy for the organism to meet its challenge, and to protect nutrients that are already stored in the body.

Other hormones, such as urocortin and prolactin, are also involved in elements of the stress response. In the brain, a number of hormones may act independently of the HPA axis and sympathetic nervous system to mediate some of the neurological and behavioural aspects of the stress response.

4

Reproductive and stress-associated behaviours: integrating differing needs

> Human beings, like all higher animals, multiply by the union of the two sexes. But neither conjugation, nor even the production of offspring, is as a rule sufficient for the maintenance of the species
>
> Bronislaw Malinowski, *Sex Culture and Myth* (1962)

4.1 Introduction

When we chance upon a person to whom we are attracted, we don't consciously consider the physiological processes suddenly active within that person. Our decision to strike up a conversation is not based on the levels of sex steroids and glucocorticoids coursing through their body, or the status of their gonads. No, we base this decision on behaviour. Is the person friendly, do they return my smile, are they looking at me or avoiding my gaze? Social and sexual behaviour are the manifestations of the underlying physiological state. In more complex organisms, such as humans, but in many other non-human species as well, behaviour becomes the most important window through which we can observe the physiological state. In species that are less neurologically complex, direct physiological interaction in the form of olfaction (pheromones), or perhaps electric fields in the case of numerous fishes are the primary methods used to consider a prospective mate.

Behaviour is the non-verbal communication used in sufficiently complex organisms to convey their intent or needs to individuals of the same species. Common behaviours help bind species into a social network. This is a

Sex, Stress and Reproductive Success, First Edition. David A. Lovejoy and Dalia Barsyte.
© 2011 John Wiley & Sons, Ltd. Published 2011 by John Wiley & Sons, Ltd.

species-specific phenomenon and the behaviour of one species is rarely recognized by another species. This is particularly true of the reproductive behaviours, although some aggressive behaviours may have meaning across a number of related species. When two or more motivational states are present, gradients and mixtures of behaviours may be present. The more complex an organism, the greater the range, mixture and nuances of the behaviours that may be present.

Behaviour reflects the underlying physiological state of the animal. If this state changes then so does the behaviour expressed. Of course this isn't always the case with humans, because we have the conscious ability to mask our internal state and express behaviours to hide our condition to others. The onset of a particular suite of behaviours is dependent upon the physiological situation the animal finds itself in. Thus, hormonal feedback mechanisms from virtually all physiological processes will act to modulate the behavioural circuits in the brain. There are distinct behaviours that are associated with a particular motivational state. In general, we can group behaviours as sexual, courtship, parental, aggressive and stress-related, for example.

The physiology of behaviour is studied in a number of different ways. One method is to administer a hormone into particular regions of the brain then observe the resulting behaviour. This can also be done by using a drug that either mimics or blocks the hormone. Frequently these studies will indicate if a hormone is administered to one part of the brain, it may elicit a particular behaviour. If, however, the same hormone is added to a different part of the brain, a different behaviour may emerge. The behaviour itself arises from a complex set of interactions among numerous brain regions, and hormones are the messengers that stimulate or inhibit these regions of the brain. One reason for this is that different complements of behaviours may emerge during a particular motivational state. For example, for many species, the care of the young may induce parental behaviour, but it may also elicit a number of aggressive behaviours in order to protect the young from potential predators. If these behaviours in animals are studied under laboratory conditions, as they frequently are, hormones can be added to the brain and the response recorded. It may be found, for example, that when a particular hormone is added to one part of the brain it causes an increase in parental behaviour, but when added to another part of the brain it induces increased aggression toward other individuals. Another method used to study behaviour involves the destruction of a small part of the brain and examination of the resulting behaviour.

Recently, there has been increasing use of specific targeting of a given gene to reduce its activity using, for example, short hairpin RNA (shRNA) or small interfering RNA (siRNA) technologies. Such studies can be used to ablate the activity of a certain group of cells, or may be used to inhibit the production of a particular hormone or its receptor.

4.1.1 Describing and defining behaviour

In the cold logic of science, behaviour is simply a unique pattern of movements that can be described in time and space. In simple organisms, behaviours may be stereotyped in that, for a given stimulus, there may be an almost identical set of movements that occurs each time the appropriate stimulus is given. Behaviours may be simply motor programmes that are activated when the need arises. In more complex organisms, the motivation for certain motor programmes may be stimulated but the programme itself will only occur after a series of additional positive stimuli are given. For example, if you are with a group of friends and you feel hungry, you may decide not to eat at that moment if none of your friends are hungry. If negative stimuli are present, then the initial programme may be attenuated. So, if your friends all decide that they do not want to eat now, you may ignore your hunger. In situations where the stimuli have both positive and negative elements a certain amount of anxiety may be present and the resulting behaviour may be a bit ambiguous. So, you may feel hungry and want to eat, but if your friends are all on diets and you do not want to upset them, you might grab a snack when they aren't watching.

Studying behaviour in complex organisms is difficult, although that doesn't stop us from doing it. We study behaviour for a variety of reasons. We might be interested to understand behaviours in a particular species. And so we describe a set of behaviours that species indulges in with the goal of learning more about that species. However, we might also study behaviour in another species with the goal of understanding more about human behaviour. For example, if we want to understand how a particular hormone or drug affects human behaviour, we would first administer it to another species, usually a rodent, then observe the resulting behaviour and establish a relationship between that behaviour and similar behaviours we see in humans. Studying behaviour in simpler organism such as rodents is easier, and there are a number of ethical concerns of administering unknown compounds to humans. But now we are faced with the challenge in determining which behaviours in rodents are analogous to those in humans. It is challenging to establish which human behaviours are present in a rodent because of the complexity of humans and the likelihood that many behaviours in humans are unique. Even if we identify motor programme behaviours in rodents that are similar to those humans, that does not mean that their motivational state is related.

Thus, while we describe a number of behavioural studies in various animals, the reader is cautioned that such findings may not be applicable to other species.

4.1.2   Sexual behaviours

A number of interacting hormone systems regulate courtship and reproductive behaviours (Figure 4.1). There are two basic types of systems that affect behaviour. These may be loosely grouped into chronic and acute effects. The onset of many of these behaviours are brought about by sex steroids but they can be attenuated by stressful stimuli. Sex steroids typically play long-term transformative roles, for example, in the neural organization of masculine or feminine behaviours. In animals undergoing a seasonal breeding period, or in oestrous females, sex steroids can induce a readiness for reproductive behaviours. However, a number of other hormone systems may provide an acute modulation of the basic reproductive or sexual state of the animals. If we were to make a generalization, then in more neurologically complex organisms, sex steroids play a decreased role in the day-to-day processing of novel sexual or reproductive stimuli. In other words, sex steroid 'priming' might tell us that we

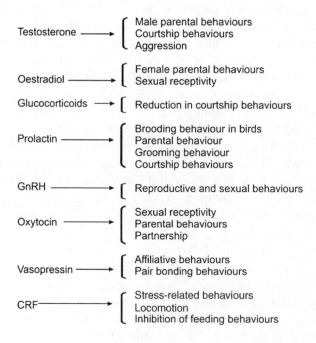

Figure 4.1   Induction of some behaviours by reproductive- and stress-associated hormones

are ready and able to have sex and reproduce, but other systems will actually allow us to engage in the activity. Males of many species, notably birds, for example, will perform an elaborate courtship display indicating their willingness and readiness to mate. However, it is only when a female indicates that the feeling is mutual that he is able to carry out his wishes.

Males and females of most species typically display different complements of behaviour. The term 'sexual dimorphism' is used to indicate aspects that differ between the sexes. It can be used to describe anatomical, physiological or behavioural differences.

The brains of males and females are different. They differ in numerous regions with respect to the numbers of cells and the numbers of connections among the cells. These differences in the brain are referred to as sexually dimorphic nuclei, where a nucleus refers to a cluster of cells. The sexually dimorphic nuclei of the forebrain are induced during early periods in development and include the preoptic nucleus, medial bed nucleus of the stria terminalis and medial amygdala (Figure 4.2).

The medial preoptic area is important for male sexual behaviour. This region is richly interconnected with other regions of the brain associated with the expression of various behaviours. In experimental animals amygdala lesions reduce the motivation to gain access to females and also reduce the male's sexual performance in the presence of females. Disruption of the medial preoptic nucleus, bed nucleus of the stria terminalis and medial amygdala interferes with male sexual behaviour. Female animals exposed to high levels of testosterone in the womb show an increase in masculine behaviours.

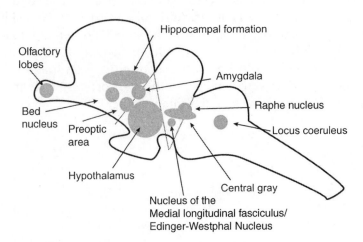

Figure 4.2 Regions of the brain associated with some of the behaviours described in this chapter

Exposure to testosterone and a decrease in oestradiol can also lead to an increase in body weight.

Sex steroids regulate male and female sexual behaviour but also singing in birds, terrritorial behaviours and spatial abilities in mammals. Sexual behaviours that are under the control of steroid hormones are not necessarily the same for all species of animals. For example, rodent behaviours are closely linked to steroid levels whereas in neurologically complex primates the behaviour is more independent. Sexual behaviours can be induced by a number of hormonal events. Lordosis is the stereotyped response that female rodents display as a receptive behaviour in the presence of males. It can be triggered by initially priming the circuits with systemic administration of oestradiol and progesterone then centrally administering oxytocin. The ventromedial hypothalamus appears to play a role in eliciting lordosis. Oestradiol infusion into this region will lower the threshold for lordosis. It also elicits sexual receptivity via the induction of oxytocin receptors when administered to the ventromedial hypothalamus of parthenogenic (female only) species of whiptail lizards. Progesterone will also induce oxytocin receptors in oestradiol-primed rats.

A number of other hormones, however, such as endorphins, gonadotrophin-releasing hormone (GnRH), dopamine and others can also influence sexual behaviour. In rats, male stereotypical sexual behaviour includes pelvic thrusting, mounting, neck gripping and ejaculation. These behaviours are potentiated by testosterone. Testosterone has an organizational effect on the brainstem and spinal cord sites that are essential for penile erection. It appears to enhance neuronal survival in these sites by the activation of neurotrophic factors. Androgens are necessary for crowing and strutting behaviour in male quails and probably most avian species. In newts, testosterone and vasotocin also regulate male sexual behaviours. Arginine vasotocin or GnRH injections will stimulate sexual behaviour in both male and female newts. This system can be inhibited by glucocorticoids or activation of the hypothalamic–pituitary–adrenal (HPA) axis. Corticosterone can inhibit androgen secretion and reproductive behaviour. In newts, corticosterone may reverse the neurophysiological effects of arginine vasotocin.

4.1.3   Aggressive behaviours

Aggressive behaviours are linked both to reproduction and stress. Aggression might be defined as a tendency to confront a novel situation rather than running away from it. In earlier chapters we discussed the 'fight or flight' response. Aggression is the manifestation of the fight response. Many aggressive behaviours are sexually dimorphic and are also linked to the sex steroids.

Castration of male rats at birth, which results in reduction in the level of circulating androgens, is associated with a reduction in inter-male aggressive behaviours as the males get older. Restoration of testosterone by injection will potentiate these behaviours. Similarly, injecting newborn female rodents with male hormones causes the them to behave more aggressively. However, castration of a male rodent more than a few days after birth does not cause a reduction in the proclivity for aggressive behaviour, suggesting that the critical period for aggressive behaviours occurs during the same critical period as required for normal masculinization. Thus, it is likely that testosterone acts to lower the threshold for fighting, thereby establishing a biological readiness for normal aggressive behaviour and facilitating the expression of aggression in appropriate social settings.

Aggression in rodents appears to involve the medial preoptic nucleus and anterior hypothalamus, medial amygdala and medial bed nucleus of the stria terminalis, where the latter is particularly important. Testosterone may exert its effects on aggression in part by modulating vasopressin receptors.

Females will also fight when provoked, indicating that the relevant circuitry is present. In the spotted hyena the females are particularly aggressive, and have been used as a model to understand the role of androgens in aggression. Female hyenas are both larger and more aggressive than males as a result of being exposed to higher concentrations of testosterone and its metabolites in the womb. Females also have external male-like genitalia and exhibit rougher play than males. Androstenedione secreted by the ovary is changed to testosterone within the placenta and directly affects the development of the fetus. Testosterone concentrations are about the same in males and females but androstenedione levels remain high in females.

4.1.4   Scent marking

Scent marking is associated with both sexual and territorial behaviours. Central infusion of vasopressin can facilitate scent marking by Syrian hamsters. The bed nucleus of the stria terminalis and medial amygdala have been implicated in this behaviour. In contrast to most species, in Syrian hamsters the females tend to be more dominant and there is greater flank marking than in males. Oestradiol, by increasing vasopressin gene expression facilitates the likelihood of flank marking behaviour. In many mammals vasopressin levels are higher in the medial amygdala, bed nucleus of the stria terminalis and lateral septum in males than in females. The amount of vasopressin expression in the bed nucleus of the stria terminalis is determined by androgens during the critical developmental period.

4.1.5   Social behaviours

Social behaviours include a number of affiliative and pair-bonding behaviours that are essential to establish the cohesiveness of a population. Voles have been utilized as a model for understanding the neuroendocrinology of social behaviours. The effects of oxytocin and vasopressin play a crucial role on the social behaviour of male and female voles. Both female and male voles injected with arginine vasopressin or oxytocin show increased social contact and partner preference compared to control groups. In contrast, males and females injected with an arginine vasopressin antagonist show little social contact. However, when the antagonist was combined with oxytocin there was an increase in social contact but not partner preference. When an oxytocin antagonist was administered there was little social contact. Administration of the oxytocin antagonist and arginine vasopressin resulted in an increase in social contact but no partner preference in both female and male voles. Partner preference was not exhibited with either of the antagonists, suggesting that both arginine vasopressin and oxytocin are required for partner preference.

There are three basic forms of the vasopressin receptor, V1aR, V1bR and V2b. The V1aR vasopressin receptor facilitates affiliation and pair-bond formation through its actions in the ventral pallidum of the ventral forebrain. The pallidum is part of a group of structures called the basal ganglia situated at the base of the forebrain that are richly connected with the cortex. The basal ganglia are associated with a range of functions including motor control and learning. The ventral pallidum is associated with reward and performance in a particular location.

In a series of interesting studies, transgenic voles with an overexpression of the V1a vasopressin receptor were created. Normally, in the presence of females male voles do not form a partner preference. However, in the group where the receptor was overexpressed, the males displayed high levels of affiliative behaviour compared to the control groups. It was also found that high levels of anxiety increased pair-bonding in male voles. Social preferences are sexually dimorphic in prairie voles. The stress of swimming and injections of corticosterone have different effects in male and female voles paired with the opposite sex. Male voles under swim stress and/or corticosterone administration facilitated partner preference for the familiar partner. However, females injected with corticosterone show no partner preference. Furthermore, when male and female voles are adrenalectomized, thus eliminating corticosterone, males did not form partner preferences whereas females did. These landmark studies provided considerable insight into understanding the neural mechanisms of pair-bonding, although it is not entirely clear how these findings relate to humans.

In general, it appears that similar mechanisms are present in humans. Do these findings reduce our own relationships to a set of hormonal interactions?

Probably not. But they did indicate that these hormones are probably involved in similar behaviours in humans.

4.1.6 Vocalization

Both reproductive and stress-related hormones have been implicated in vocalization. In many situations it is not clear what the purpose of a vocalization of a given species is. It may be associated with courtship, aggression or perhaps a number of other aspects. The neuroendocrine regulation of vocalizations has been probably the most well studied in avian species. In the zebra finch, testosterone potentiates the recognition of a conspecific song. Songbirds sing during the spring when the high vocal centre (HVC) nucleus of the brain enlarges as testosterone concentration becomes elevated. Testosterone and testosterone conversion to oestradiol induce interneuron communication in the neuronal circuitry required for song production. Song production may also involve vasotocin expression facilitated by testosterone. In male starlings, testosterone is high in the spring when calling is used in part to attract mates. Many of the brain nuclei that regulate song contain high densities of adrenergic receptors and testosterone can regulate the density of α_2-adrenergic receptors in the brain, suggesting that noradrenaline is also a key neurotransmitter associated with singing.

In frogs, arginine vasotocin potentiates the motivation to call. In green tree frogs, arginine vasotocin can increase the probability of calling but does not affect either the latency period or how often individuals were observed calling. Corticosterone can inhibit calling in arginine vasotocin-treated males only at high doses. Lower concentrations do not appear to affect calling behaviour. Corticosterone also reduces androgen levels. Endogenous androgens levels are inversely correlated with the latency to begin calling, suggesting a potentiating effect on calling by androgens. Plasma levels of dihydroxytestosterone are significantly increased in male southern leopard frogs when exposed to a prerecorded chorus of conspecifics. Vocalization in the African clawed frog is sexually dimorphic. The elaborate singing of the males is induced by testosterone. Testosterone induces cell proliferation in the larynx in males and prevents the muscle fibre loss that occurs naturally in females. A number of neural sites also regulated by testosterone are associated with the control and regulation of the song. A vocal motor pattern generator in the tegmentum, sensory region of the thalamus, the preoptic region and the striatum are involved.

An analogous situation occurs in fishes that utilize electric fields for perception. The brown ghostfish, for example, employs electric 'chirps' as a mating call. These calls are regulated in part by the pre-pacemaker nuclei of the hypothalamus, which is intensely innervated by corticotrophin-releasing factor (CRF)-immunoreactive fibres, suggesting a role for CRF in the chirping behaviour in this species.

4.1.7   Parental behaviour

A number of parental behaviours are similar to affiliative and pair-bonding behaviours. Both oxytocin and vasopressin have been implicated in the regulation of these behaviours. Central infusion of vasopressin has been shown to increase parental behaviour in prairie voles. Testosterone concentrations modulate vasopressin receptors in the brain. Eliminating testosterone and lowering central vasopressin concentrations will reduce male parental behaviour in voles. Castration also reduces vasopressin expression in voles. Oxytocin can also elicit maternal behaviour. High oxytocin leads to maternal attachment whereas low oxytocin can lead to withdrawal. Central, but not peripheral infusion of oxytocin will elicit partnership in prairie voles. Oxytocin can also stimulate female aggression. Centrally infused oxytocin will also stimulate parental care in male mice and may also be elevated in the expectant fathers.

In the monogamous prairie vole, oxytocin receptors are more numerous in several brain regions including the bed nucleus of the stria terminalis compared to the non-monogamous mountain vole. There is greater male parental care among the monogamous species. The greater amount of time they spend with pups is correlated with a greater the number of oxytocin receptors. However, high concentrations of corticosterone, resulting from prolonged stress, will reduce attachment behaviours in voles. Oestradiol will increase oxytocin-expressing cells in the bed nucleus of the stria terminalis. This region appears to play a role in social behaviours and formation of partner preference. Oxytocin has been implicated in a number of behaviours associated with sexual arousal, pair-bonding and sexual satisfaction. Oxytocin is also involved with parturition and lactation. It can either elicit sexual receptivity or parental behaviour depending upon its site of action.

A role for prolactin in parental behaviour in both sexes has been documented in a number of mammalian orders including primates, ungulates (hoofed animals), rodents, and lagomorphs (rabbits, hares and pikas), and in birds. Several brain regions may contain prolactin receptors through which prolactin mediates its influence on behaviour and pituitary hormone secretion. Infusions of prolactin into the preoptic region of oestrogen-primed, ovariectomized virgin rats induces maternal behaviour. The behavioural effects of prolactin are probably due to either the prolactin generated by the anterior pituitary, which is then released in to the brain via uptake across the choroid plexus and possibly circumventricular organs, or due to prolactin expressed and released in specific regions within the brain.

Prolactin is found in a number of regions of the brain associated with behaviour, including the bed nucleus of the stria terminalis, medial preoptic area, prefrontal cortex, septal nucleus dorsal hippocampus, midline thalamic nucleus, central grey, dorsal raphe nuclei, locus coeruleus, nucleus of the solitary tract and medial amygdala. Prolactin receptors are present in the preoptic area, ventromedial hypothalamus, bed nucleus of the stria terminalis, suprachiasmatic nucleus and central grey. Suckling increases prolactin secretion in both mother and offspring from both the pituitary and hypothalamus. Decrease in attachment between a rat pup and mother will decrease prolactin and increase glucocorticoids in the pup. Prolactin administered into the forebrain of female rats after being primed with oestradiol will induce the rat to respond with maternal behaviour such as pup retrieval and licking. Prolactin administration into the medial preoptic region will also elicit this behaviour, whereas lesions in this region will destroy this behaviour.

Prolactin also elicits brooding behaviour in both sexes of birds. In turkeys, prolactin infusions into the brain can increase brooding and nest-making. Nest deprivation decreases prolactin. In doves, prolactin increases food regurgitation to neonates. In ring doves, prolactin concentrations in males and females increase during the period of enhanced parental care. At the beginning of the rearing period, the female does most of the regurgitation, whereas toward the end of the rearing period the male does more of the regurgitation.

4.1.8   Locomotion

A CRF-mediated role in locomotor activity was established shortly after the discovery of the peptide in 1981, and has received much attention. However, amphibians and fishes may be particularly appropriate models for investigating the mechanisms of CRF on the neural circuitry. CRF can stimulate locomotory behaviours in newts and salamanders and the peptide's actions can be modified by the presence of opioids. In newts, discharges of neurons of the raphe region associated with walking and swimming behaviours can be modified by injections of CRF into the lateral ventricle of the forebrain.

4.2   An integrated approach to behavioural modulation

Learning is essential to survival. We want to repeat the behaviours that lead to benefits and avoid the situations that threaten us. Learning occurs in most organisms to varying degrees. But how we learn, and what we perceive as being significant can vary according to our motivational state. The memory and encoding of emotions and stimulus-response events are channelled through a common set of neurological structures that also receive input from neuroendocrine factors associated with stress, reproduction, diuresis and feeding. The level of integration, and the strength of signal between environmental stimuli and physiological responses of the *milieu intérieur* dictates how the memory will be encoded and how the organism will behave in a similar situation the next time a similar stimulus is perceived.

Because reproductive, avoidance and aggressive behaviours possess a number of similar features in vertebrates, we may postulate then that the neurological basis of these behaviours are intrinsically linked at some part in the brain. For example, head-bobbing behaviour in some lizards can signify aggression between males or courtship between males and females. Immobility in rats is one of the characteristics of receptive behaviour and of fear or avoidance behaviour. Central administration of CRF-like peptides can induce a period of immobility followed by an attenuation of exploratory behaviour. However, electrical stimulation of the midbrain central grey will also facilitate lordosis and lesions in the central grey will disrupt it. GnRH infusions into the central grey will potentiate lordosis behaviour, whereas immunoneutralization will inhibit the response. Infusions of the GABA receptor antagonist 'bicuculline' or antisense oligonucleotides to GABA-synthesizing enzymes into the central grey reversibly reduce typical female sexual behaviours such as proceptivity and lordosis in the rat. In the central grey, anxiolytic drugs such as diazepam can not only facilitate lordosis, but also reduce the fear response in rats, suggesting that both mechanisms are linked. Indeed, a fear response must be inhibited before sexual receptivity can occur.

The interaction of such circuits may be associated with urotensin-I/ urocortin and GnRH-II in the midbrain. The nuclei in which they are found (medial longitudinal fasciculus or Edinger–Westphal (EW)) may, therefore, regulate motor behaviours associated with reproductive activity and stress. Classically, studies suggest that cells of the central grey synapse onto neurons within the reticular formation of the medulla, which subsequently send their axons into the spinal cord to control the motor neurons. However, the role of the medial longitudinal fasciculus and EW nuclei in the regulation of these behaviours has not been investigated. For example, the nucleus of the medial longitudinal fasciculus extends fibres to all levels of the spinal cord. The presence of urotensin-I in the medial longitudinal fasciculus suggests that it plays a role in the modulation of spinal cord motor circuits.

4.3 Stress and the modulation of learning and behaviour

Adrenal stress hormones can act to facilitate the storage of a memory if they are given within a particular time interval after the event. Although certain levels tend to facilitate the memory, higher concentrations can attenuate the memory experience. A number of psychiatric disorders such as depression and anxiety-related (mood) disorder have been associated with the dysregulation of the HPA axis.

4.3.1 Neurohormonal pathways

Adrenaline and glucocorticoids can improve the retention of memory. Glucocorticoids can pass readily across the blood–brain barrier, and can gain access to the amygdala. Adrenaline, on the other hand, cannot pass through the blood–brain barrier. One possibility is that peripheral adrenaline activates neurons of the nucleus of the solitary tract in the brainstem that subsequently project to the locus coeruleus. The locus coeruleus contains the principal noradrenaline projections for the forebrain and, in addition, projects directly to parts of the amygdala. During stress, locus coeruleus neurons discharge at a higher rate, and there is an increase in noradrenaline in the amygdala during these periods, suggesting that this circuit is responsible for the β-adrenergic-mediated pathway. During emotional arousal, 5HT, dopamine, noradrenaline and acetylcholine are all released centrally in a much higher concentration than under normal circumstances. This suggests that there is an activation of these circuits during emotionally charged events. However, only β-adrenergic and muscarinic acetylcholine receptor antagonists can block the emotion-potentiating effects on memory.

The pathway mediating the cholinergic responses is less clear. Cholinergic projections from the basal forebrain to the cerebral cortex have been implicated in cortical plasticity. The substantia innominata is a heterogeneous region situated under the anterior commissure in the forebrain, and includes the nucleus basalis. The nucleus basalis is the single major source of cholinergic innervation of the entire cerebral cortex. About 90% of the cholinergic neurons of the nucleus basalis project to widespread regions of the cerebral cortex. Thus, the nucleus basalis may be analogous to the raphe nuclei and locus coeruleus that supply the majority of the serotonergic and noradrenergic nerves, respectively, to the cortex. Afferents to the substantia innominata/nucleus basalis regions arise mainly from the amygdala, insular and temporal cortex and from the pyriform and entorhinal cortices. The basolateral amygdala is reciprocally connected to the substantial innominata, suggesting that it is anatomically situated to receive and modulate acetylcholinergic pathways.

Lesioning of the basolateral amygdala, but not the central nucleus of the amygdala, can block the memory-enhancing effects of adrenal glucocorticoids. A similar situation occurs with the disruption of the stria terminalis, which is a major output pathway of the amygdala. In emotionally arousing situations, when under the influence of peripheral stress hormones, the amygdala facilitates the consolidation of long-term memory in other brain regions where memories are actually stored (e.g. the neocortex).

The basolateral amygdala facilitates long-term potentiation of the perforant path input to the dentate gyrus. This pathway integrates brain cells of the entorhinal cortex with all parts of the hippocampus. The entorhinal cortex is pivotal for the integration of memory and navigation and provides an interface between the hippocampus and cortex. The dentate gyrus, which is part of the hippocampal formation, is implicated in memory formation and the formation of new brain cells. The basolateral amygdala does not project directly to the medial septum and locus coeruleus which are the main sources of noradrenaline, thus if the basolateral amygdala acts to potentiate the consolidation of memories in the hippocampal regions by adrenergic and acetylcholinergic pathways, then it may be an indirect mechanism.

4.3.2   CRF, glucocorticoids and hippocampal function

The hippocampal function is influenced by glucocorticoids through the modulation of neuronal excitability. Glucocorticoids can affect hippocampal-dependent behaviour such as spatial memory. This mechanism may also play a role in ageing, as some studies indicate that glucocorticoid levels are correlated with a reduction of hippocampal volume and memory deficits in aged rats. The hippocampus, in turn, has an inhibitory action on the HPA axis. Thus, as a function of ageing or chronic HPA activation, the inhibitory actions on the HPA axis become attenuated, leading to enhanced HPA activation and a greater potential for mental dysfunction.

The organism's response to stress may impair learning and memory in adulthood. This is probably caused by the loss of neurons in the hippocampus. For example, CRF administered to the brains of 10-day-old rats for several months leads to a 17% decrease in hippocampal neurons. The loss was most pronounced in the CA3 area of the hippocampus. Twelve-month-old rats have more neuronal loss than eight-month-old rats. A decline in the performance in the water maze, and less time investigating novel objects occurred in the CRF-treated animals, indicating a deficit in some mental functions.

Glucocorticoids have affinity for both the type I and type II receptors, whereas mineralocorticoids only have affinity for the type I receptors. In aldosterone-sensitive tissues, such as the kidney, glucocorticoid bound to the type I receptor is enzymatically digested by 11-hydroxysteroid dehydrogenase, such that the confounding glucocorticoid signal is removed. In many regions of the brain, however, the type I receptor is not coupled to the enzyme and, therefore, glucocorticoids signal through both receptors. Glucocorticoids have a tenfold higher affinity for the type I receptor than for the type II receptor and, therefore, will be selectively bound to the type I receptor in the presence of mineralocorticoids. The type II receptor is found in most cell types of the body, whereas the type I receptor is only found in brain, the epithelial cells of the kidney, colon and exocrine glands and reflects the role the mineralocorticoids play in sodium and water readsorption.

In the brain, the type I receptor is found only in the hippocampus, septum and amygdala. Both type I and II receptors are, therefore, found in the hippocampus. Low glucocorticoids activate the type II receptors and high concentrations of the steroid activate both receptors. Low concentrations of glucocorticoids will excite hippocampal neurons whereas high concentrations will decrease neuronal excitation. Activation of the type I receptor has been linked to long-term potentiation, whereas activation of the type II receptor appears to favour long-term depression of the neuronal activity. However, the molecular actions of the glucocorticoids are complex and little is known about the interaction of the type I and type II receptors when both are activated. Glucocorticoids can induce transcription through the ligand-activated receptor by binding to glucocorticoid responsive elements (GREs) in the promoters of genes that they regulate. However, they can also repress transcription by binding negative GREs. The activated type II receptor can also bind other transcriptional factors such as AP-1 or NFκb to suppress their action or can facilitate the action of STAT5. Some studies also suggest that the type I and II receptors can heterodimerize to exert a distinct complement of actions.

4.4 Summary

Regardless of the organism, the onset of new behaviours and the modification of established behaviours require both learning and memory. A stimulus to an organism may be described as being positive, negative or neutral. Positive stimuli act to promote behaviours to facilitate interaction with the stimulus whereas negative stimuli act to induce behaviours that will avoid future interaction of the stimulus. Neutral stimuli are those that do not have a behaviour-modifying ability. In vertebrates, regions of the brain act to induce a physiological reward in positive stimuli whereas fear centres promote the opposite effect in the transduction of negative stimuli. The amygdala is a key region of the brain that integrates sensory and associative information to regulate the onset of both reward- and fear-potentiated behaviours. It may also play a role in memory consolidation in conjunction with the hippocampus and other sites, before the memory is stored. Many behaviours that are sexually dimorphic, including courtship, aggression and sexual behaviour, are regulated by the sex steroid milieu of the organism. The development of the brain circuitry of these behaviours appears to be achieved in utero or perinatally in response to changes in circulating sex steroids. In the adult, sex steroids may act to facilitate or inhibit a number of these behaviours by interaction with vasopressin and oxytocin-sensitive regions of the brain. In mammals, emotional arousal associated with activation of the HPA axis can enhance some behaviours but depress the incidence of others.

5

Animals under strain: life is stressful

Let those whom Nature hath not made for store, Harsh, featureless, and rude, barrenly perish:

William Shakespeare, *Sonnets* (1609)

5.1 Introduction

The total stress that an organism is subject to during its lifetime is profound. Once, many years ago, I visited a town in northern British Columbia in Canada, where you can watch salmon migrating up river to spawn as you walk along the side of the river on the relative safety of a concrete sidewalk. While I was watching these fish jump in incredible acrobatic displays as they struggled to swim up the cascading waters, I saw a person ahead of me with a T-shirt that read 'You're born, you spawn, you die'. For the salmon, that was clearly the case. I had hoped, however, there was a bit more to life than that.

As we discussed in Chapter 1, if there is no reproduction, then life as we know it could not have evolved. Life is fraught with danger, and we must survive these challenges to procreate. Similar to the salmon's gritty determination to overcome the waterfall to find the spawning grounds, all animal species will meet the challenges of life's waterfalls to reproduce. Stress and reproduction are intrinsically linked in life. We and all other species have evolved to cope with these stressors. But reproduction is notoriously sensitive to stress. It has to be. The progeny must be reared in the safest possible conditions. We must be aware that when a challenge threatens our own survival and the fitness of our young. Organism's have evolved numerous ways of perceiving and adapting to threats.

However, there needs to be a trade-off between reproductive capacity, on the one hand, and the ability to ward off stressful challenges on the other. A compromise must be struck in each species to allow sufficient reproduction while protecting self and progeny from threats. Moreover, stress is a double-edged sword. A small amount of stress may increase reproductive potential but too much stress can decrease reproductive potential. How much stress is too much? This varies between individuals, populations, species and the niches they find themselves in. What we perceive as stress is not necessarily what unconscious physiological systems of the body perceive as stress. Laying on the beach under full sun might seem pleasant, but your body is being bombarded by ultraviolet radiation and may be in danger of overheating, and so a number of defence systems become activated to keep you alive. These are the everyday stresses, which organisms have evolved to adapt to. But all animals have their own unique set of stressors. Competitors, prey, parasites and predators all evolve their own adaptations and counter-adaptations to contend with constantly changing selection pressures. Indeed, these stressors may have been the driving force for sexual reproduction in the first place.

5.2   Changing environments and stress bottlenecks

Life on our planet is in relative harmony and interacts with the existing geological and meteorological processes present. But the planet is constantly changing. There are climatic processes such as hurricanes and tornados, temperature extremes and geological events such as earthquakes, tsunamis and volcanoes. There may be cosmological events such as impacts of asteroids, meteorites and comets, or perhaps changes in solar activity that can affect life on earth. Continents drift around the planet affecting the weather, the amount of coastline and the formation of geological features such as mountains and prairies. During the Permian period, some 250 million years ago, the continents came together, forming one supercontinent. As a result of the loss of coastline and shallow water regions, combined with climatic changes, it has been estimated that as much as 80% of species became extinct at that period. Geological processes take a long time, in the order of millions of years, and the changes that led to the demise of these species were subtle and gradual. But with each change, each time the land changed, each geological or meteorological process that occurred as a function of the changing continental pattern affected the structure of the ecosystems locally before becoming more widespread.

Changes in the ecosystems affect which species become dominant and vice versa, and this can lead to changes in the competitive ability of some species and then they become lost. The makeup of the ecosystem can also affect the local climatic conditions by promoting or inhibiting moisture, oxygen, carbon dioxide and other gases. This, in turn, affects the composition of the ecosystem and the evolution of species that appear during this period. In other words, there is a dynamic interaction between the nature of life, the climate and the geological processes of the planet. Animals evolve in this changing environment.

Species are continuously becoming extinct in this changing environment, even without the help of humans. This is a process that began when the first species evolved. Since the beginning of life, far more species have become extinct than the total number of species living on the planet today. There is nothing inherently 'bad' about a species becoming extinct any more than there is of an individual dying of old age. It is a natural consequence of life. However, these environmental changes, regardless of how they originated, have placed a stress on the species. We might think of these environmental challenges that species need to survive through as 'stress bottlenecks'.

5.3   Environmental stress-inducing factors

There have been many stress challenges in the past that have arisen from a multitude of sources. For simplicity, we might think of these stressors as being divided into two categories: abiotic or non-biological stressors and biotic or biological stressors.

5.3.1   Abiotic stressors

Abiotic stressors include changes in temperature, water, gases as well as the mechanical effects of planetary meteorological and geological actions described above. Temperature is a particularly good example of an abiotic stressor. Right now, global temperature change is in the forefront of public consciousness. Numerous studies indicate that the planet is generally becoming warmer. Whether or not this has occurred because of human actions on the planet and whether this will continue has not been resolved. However, one thing is clear. The mean surface temperature of the planet has changed many times in the past. Heat is a form of energy. When temperatures are at sufficiently high enough levels, biochemical reactions required to sustain life occur more efficiently. If temperatures are too low, then these reactions are less efficient, and the organisms level of activity falls off. The fluidity of cell membranes, the viscosity of biological fluids and even the structure of proteins and enzymes are affected by temperature. Temperature is a major factor in the determination of a metabolic rate of an animal. The metabolic rate dictates the rate at which an animal must gather food. The amount of energy available to the animal, after metabolic processes have been dealt with, defines the limits available for the reproductive strategy used (Figure 5.1).

The vast majority of animals on the planet today, and in the past, are those whose body temperature changes with ambient temperature changes. We call those animals 'poikilothermic', meaning variable. Animals whose body temperature remains more or less constant during changes in environmental conditions are called 'homeothermic'. Poikilothermic animals have the advantage in that they have a lower metabolic rate and require less energy to carry out their life. They have the disadvantage in that they are much more dependent on the ambient temperature to survive. In insects, for example, the amount of power available for wing muscles is directly related to the ambient temperature. Within certain limits, homeothermic animals are more independent from ambient conditions. With their higher metabolic rate, they generate enough heat to sustain life processes in colder ambient temperatures. The trade-off is that in order to maintain this metabolic rate, they require greater amounts of energy. This means that more food is required and, therefore, the amount of energy required for food gathering is greater. Mammals and birds are homeothermic. Almost all other species are poikilothermic.

Animals can also be classified on the basis of their ability to thermoregulate, or rather to maintain a relatively constant body temperature. This can be achieved physiologically or by behavioural actions. Larger insects, such as bumblebees, or some fishes, such as tuna, are examples of this as a higher body temperature can be maintained during use of flight or swimming muscles, respectively.

The physiological strategy used to deal with ambient temperature stressors places a number of constraints on the reproductive strategy of the organism. A poikilothermic organism's energy reserves are higher when temperatures are high, and lower when the temperatures drop below a critical level. This places a severe handicap their ability to care for progeny. During the cooler nighttime temperatures, the animal is at a disadvantage due to the limited energy budget. When the animal is active, the collective metabolic activity generates a certain amount of heat. However, as the ambient temperature falls, then the animal must generate more heat by increasing activity to offset the temperature differential between what is required for normal biological functions and the ambient temperature. In order to do so, the animal will require greater reserves of food and to spend more time feeding, allowing less time for the progeny.

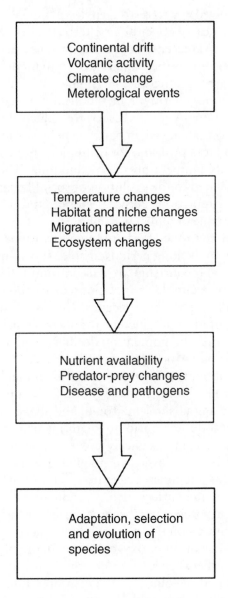

Figure 5.1 Relationship between environmental biotic and abiotic factors

Clearly, this cannot happen. The progeny, therefore, need to be self-sufficient. In almost all poikilothermic animals there is no or little parental investment in the juveniles. In homeothermic animals we see much greater levels of parental investment.

Homeothermy evolved in a terrestrial environment. Water is a strong buffer of temperature extremes so once the first terrestrial organisms evolved they were subjected to much greater temperature extremes. Some researchers have suggested that temperature is the primary determinant of a species' range. The first rudiments of a homeothermic physiology probably evolved in early reptiles. However, homeothermy is metabolically expensive and additional physiological adaptations were necessary to reduce the metabolic cost of homeothermy. This was achieved, in part, by increased insulation in the form of hair, feathers and fat deposition, for example. Another method to achieve insulation against cold temperatures is by huddling. The acquisition of this trait was probably related to the evolution of social behaviour and may have started out through interactions between mother and offspring.

Let's return to the issue of global warming. Species living in the Arctic may be particularly sensitive to the effects of global warming. Here, species have evolved and adapted to the long periods of cold, snow and frozen seas. Their life history, breeding and courtship are based around these conditions in the different seasons. For example, populations of caribou and reindeer have fallen worldwide over the past three decades. The loss of caribou could remove an important source of meat and income for northern people as well as affect their traditional culture. The population decline could also affect plants and invertebrates that rely on caribou grazing and faeces for access to nutrients. For migratory Arctic caribou, the apparent effects of climate change have had a unique impact. Warmer summers have boosted the number of insects and parasites, leading the animals to run around and shake themselves instead of feeding. If the caribou do not put on enough fat before winter, they won't survive and the females cannot conceive. Warmer winters have led to more freezing rain, which forms a layer over the lichens that the animals eat. So, something as simple as a temperature extreme can have numerous effects on animals that will ultimately affect their reproductive fitness. If the warming trend continues, some species may become extinct if they cannot adapt.

Like most other animals on this planet, humans need oxygen to breathe. In the years before the formation of oxygen on earth, organisms used anaerobic respiration. Symbiosis between an aerobic organism and an anaerobic organism was the first step in the evolution of oxygen-breathing organisms. Oxygen levels on this planet have changed a number of times throughout its history, and species have had to adapt to these changing conditions. Oxygen can accumulate in the air only if there is an imbalance between the amount of oxygen produced by photosynthesis and the amount consumed by respiration and various geological and geochemical processes. In the Carboniferous Period, for example, oxygen levels may have been higher than they are today because of the vast forests present during that time. Animal species that evolved during this period were selected, in part, for the higher oxygen levels. It has been suggested that these higher oxygen levels, perhaps as much as 35% higher than today's, could have led to a certain amount of gigantism in some animals. A notable example is the giant dragonfly, *Meganeura*, which had a wingspan of about 75 cm. However, these high oxygen levels were not stable and eventually the levels declined. Curiously, some studies indicate that oxygen levels declined to present values around 65 million years ago, around the time of the dinosaur extinction.

There are two physiological aspects related to the concentration of ambient oxygen. The higher the metabolic rate, the more oxygen is required. Higher metabolic rates require greater energy production, which is related to the amount of oxygen available to oxidize nutrients. In low atmospheric oxygen conditions, metabolic rate is limited by the availability of the ambient oxygen. In higher concentrations of ambient oxygen, metabolic rates have the potential to become higher. The other physiological aspect to consider is that if the limit of metabolic rate is dependent upon the availability of ambient oxygen, then the ability of the organism to extract oxygen from its environment can place limits on its metabolic rate.

Okay, fair enough. But how does this affect reproduction? As we alluded to in Chapter 1, reproduction is expensive in energy terms. Therefore, the investment in current reproductive needs must be traded off against the organism's future reproductive potential and their survival probability. Metabolic rate is related to lifespan. In general, the higher the metabolic rate the lower the lifespan, and vice versa. Associated with this is that animals with longer gestation periods tend to have lower metabolic rates and those with shorter gestation periods have higher metabolic rates. A longer gestation period means that the embryo or fetus is subject to a greater probability of a stressful challenge. This means it must be protected for a longer period of time, and in the case of parental investment, a longer period in which the mother needs to divert energy and resources to the care of the young.

There is a second issue. Oxygen, paradoxically, can be toxic. Reactive oxygen species (ROS) are by-products of the various oxidation–reduction reactions present in the organism. Many of these ROS perform useful tasks in the organism and have been implicated in immune defence, programmed cell death (apoptosis) and stress acclimation, but at high concentrations they can be highly injurious to the cell. If allowed to continue, this oxidative stress will result in continued damage to DNA, RNA and proteins in the cells. Over time, ROS can wear down an organism and reduce its lifespan. An excess of these free radicals and accumulation of their affects through life has been used to explain some elements of the ageing process. This is known as the free radical theory of ageing.

Mitochondria, the organelles in the cell responsible for oxygen-associated energy production, are responsible for the majority of the free radicals produced by respiration, and, therefore, this theory has more recently been modified into the 'mitochondrial theory of ageing', which also predicts that the higher the metabolic rate, the higher the production of free radicals. To a certain degree, this appears to be true. Animals with low metabolic rates tend to live longer than animals with high metabolic rates, although some species appear to be anomalous in this regard. For example, a pigeon has the same metabolic rate as a rat, yet can live ten years longer. However, studies on free radical production in the pigeon mitochondria indicate that respiration is far more efficient and only produces a fraction of the amount of free radicals produced by rat mitochondria. This appears to be a modification that birds have developed to accommodate the large amounts of oxygen required to power the flight muscles. So, if an organism has a particularly high metabolic rate, then the accumulation of these free radicals may play a role in an early death.

The organism needs to defend against this oxidative stress using a number of 'antioxidant' mechanisms. These are a set of molecules and proteins that are produced by the body to 'mop up' free radicals and turn them into a molecular form that is less injurious or perhaps even beneficial to the organism. There are several enzymes, such as superoxide dismutase, catalase and glutathione peroxidase, that act as antioxidants, plus proteins such as the metallothioneins, and a host of other molecules including melatonin, vitamin E, vitamin A and the carotenoids, and glutathione to name a few. This will be discussed in more detail in Chapter 9.

Several studies have implicated ROS in the 'cost' of reproduction. In the fruit fly, for example, the stimulation of egg production was associated with a decreased resistance to an oxidative stress-inducing compound. Further studies on the zebra finch showed that the more breeding activity there was in these animals, indicated by the number of eggs laid by a breeding pair, the less resistance to oxidative stress could be measured. This relationship could be ablated when antioxidants in the form of carotenoids were added to their drinking water. Dietary carotenoids, besides acting as antioxidants, are pigments and can produce colouration in the yellow-red region that is found in birds, reptiles and fishes. This colouration is frequently used as a secondary sex characteristic, usually in males, to indicate sexual robustness and dominance. Some studies, though not all, indicate that increased oxidative stress reduces the intensity of the colouration of these secondary sex characteristics. Physiologically, this is thought to involve the use of these carotenoids for antioxidant function, rather than for the pigment function. The resource is being favoured to protect the individual first and compromise the need for reproduction at this moment.

Why should this happen? ROS are produced by a number of cell types in the immune system to destroy invaders such as bacteria, parasites and other disease-causing organisms. This is an evolutionary ancient mechanism of defence, and is used by both plants and animals. ROS are non-specific with respect to their reactions. If there is a sufficient activation of the immune system by pathogenic invaders, then there will be an overproduction of ROS. Carotenoids, reacting with ROS will be unavailable for incorporation into tissues for pigmentation. This phenomenon has led to the theory that females of these species prefer the males with the most intense colouration because they are the healthiest.

Other studies have linked an increase of testosterone in males with increased colouration but with some depressed immune functions. Ultimately, however, if this is the reason for the female selecting the male, it has to relate to a number of attributes, including the male's fertilizing ability, as well as the complement of genes it can pass to increase survival and success of the offspring. There is considerable evidence that ROS can adversely affect sperm function (see Chapter 9), however there are few studies that show a strong relationship between colouration, ROS activity, immune function and sperm viability.

5.3.2 Biotic stressors

Biotic stressors are factors that are produced within the organisms themselves and include stressors such as nutrient availability disease, overcrowding, dominance hierarchies, competition among individuals, species and populations.

Food comes and goes. At times, food will be plentiful and at other times, not so. After over a half billion years of metazoan evolution, a number of physiological elements have evolved to accommodate this variability. As we have pointed out in earlier chapters, the goal of an organism is to survive long enough to procreate. Depending on the region of the planet in which the organism is living, food may be available seasonally. Moreover, the amount of food available may vary within the season or in the same season in different years for the reasons discussed above. When food is plentiful and all other stressors are minimal, then the animal undergoes normal sexual maturation and produces the expected progeny. However, at other times, food may not be available. In this case, the animal may undergo a stress response in which sex and reproduction is inhibited, but other physiological systems become activated to enhance the individual survival. This phenomenon is termed 'caloric restriction'.

A number of studies have indicated that calorie-restricted diets may extend the lifespan of a number of organisms. Naturally, as we baby-boomers are approaching our senior years, we are particularly interested in anything that will extend lifespan! In calorie-restricted diets, calories are reduced by 30–40%. In general, the composition of nutrients in these studies was balanced, although it is possible that in some species the ratio of different nutrients, for example protein and carbohydrates, and not caloric intake per se may contribute to an anti-ageing effect. The calorie restriction in these studies was considerably less than a starvation diet, in which calorie intake is less than 50% of the normal diet. Some studies indicate that lifespan can be increased 30–50% in most animals studied.

Moreover, caloric restriction in rats not only increases lifespan, it also delays the effects of ageing. This sounds too good to be true, right? There is a downside, naturally. The overall effect of calorie restriction is to increase stress resistance. Resistance to oxidative stress increases, particularly in the brain, heart and skeletal muscle, which are required during fight-or-flight responses, and where oxidative stress can do the most damage. In other words, metabolism is shunted away from sex and reproduction to protect the body from potential damage.

Depending on the species and the amount of caloric restriction, an individual may not remain in an active stress response situation for the entire restriction period, but will be able to mount a stress response much quicker in the face of an appropriate challenge. Now, before we all decide to give up sex and reduce our calorie intake in order to live longer, it is not clear that caloric restriction late in life will have an appreciable affect on increasing lifespan. In evolutionary terms, the mechanism appears to be in place to protect the individual during times of decreased food availability, and to hold off reproduction until some time in the future when the conditions are better.

We have been discussing the concept of food as if it was passive, as if it lovingly throws itself toward the species that is trying to survive; to sacrifice itself for the good of another species. Of course, no such thing happens. Before food becomes food, it is a living organism and it, too, needs to survive long enough to reproduce. It doesn't matter if it is a plant, fungus or animal, it will defend itself to the best of its ability. Let's consider autotrophs (plants and fungi) for the moment. Besides the myriad of mechanical methods to defend themselves in the form of hairs, spines and bark, for example, there are a number of chemicals that may be produced that have a noxious or physiologically modifying effect on the foraging species. Here, it is worth mentioning the nutrient–toxin relationship. Any nutrient in a sufficiently high dose can act as a toxin and be injurious to the ingesting organism. By the same token, any toxin in sufficiently low dose can act as a nutrient. In low enough concentrations the organism can metabolize the toxin and use it as food. Most food sources will contain a combination of nutrients and other molecules that may be considered toxic. At times, it may be necessary to ingest these plants if other food sources are scarce.

A number of plants have evolved compounds that act as hormone mimics in animals. It is unclear why this has occurred, although one theory is that these compounds can act as contraceptives on grazing species to limit their proliferation. The most intensely studied of the hormone mimics are those that possess steroid-like activity. Because of their lipophilic activity, these hormones are stored in adipose tissue and can bioaccumulate. Moreover, many readily pass through the blood–brain barrier, allowing for potential to affect all neural and neuroendocrine processes. Steroid-synthesizing pathways are well established among a number of plant lineages. A number of species can manufacture phytoecdysteroids that mimic the natural moulting hormones of insects. They are stored in the leaves, and can disrupt the normal development of the insects that feed on them. In sedges, a compound with the same structure as juvenile hormone in insects is produced. Insects that feed on these plants show a disruption of growth and reproductive ability. Plants can also make a compound that interferes with juvenile hormone production in insects. This substance, precocene, is processed by the insect's brain and kills the cells. As a result, juvenile hormone is not produced.

This effect is not limited to insects. Phytoestrogen is a generic term referring to chemicals in plants that have oestrogen-like activity. The most well characterized of the phytoestrogens are the isoflavonoids. These are almost entirely restricted to the Leguminosae/Fabaceae family (peas, beans, clover). Red clover, for example, produces alkaloids that mimic oestrogen and cattle eating too much red clover may suffer from reproductive disturbances.

Hundreds of different isoflavonoids have been identified. Many have physiologically active oestrogen-like effects. The isoflavonoids daidzein and coumestrol from clover have been shown to have sufficient activity to modulate the reproductive activity of grazing animals. This may have developed through the evolutionary coexistence of herbivores and plants where there was a need to reduce grazing pressure in order for the plants to survive.

Although numerous studies have shown that high concentrations of isoflavonoids can modify a number of hormone and enzyme systems in animals, the effects on humans are controversial. Some early studies suggested that excessive ingestion of soy isoflavones is associated with decreased plasma testosterone and androstenedione levels in human males, but more recent studies indicate that this may not always be the case. Similarly, an increased intake of isoflavones has been implicated in higher bone mineral density in post-menopausal women. This does not seem to occur in pre-menopausal women and may indeed be associated with a decrease in bone density. Moreover, some authors have postulated there may be an increased risk of cardiovascular disease or aberrant menstrual cycles in women ingesting high isoflavone diets. In contrast, epidemiological studies have suggested that isoflavones are associated with reduced breast cancer risk in women and reduced prostate cancer risk in men. For this reason, isoflavone supplements have been promoted as a dietary supplement to reduce some cancers and to normalize reproductive function. However, establishing a clear cause and effect between isoflavone diets and human physiological changes has proven problematic. Cultural and genetic differences among individuals have been cited as confounding factors. Unfortunately, there are few good controlled studies on humans to address these issues, and many of the health concerns regarding enriched isoflavone diets are the result of extrapolation of rodent studies. Rodent physiology and toxicology studies do not translate well to human studies. Nevertheless, there is sufficient concern, as a result of the studies, to suggest that limits be placed on the amount of isoflavones ingested daily.

5.4 Migration as part of a life strategy

Have you ever moved away from home? Or sold your house and bought a new one? Or perhaps you have moved to a new city, or country? The process of moving can be very stressful. Some studies suggest that moving to a new location, in terms of stress, is second only to the death of a loved one. A number of species migrate as part of their normal life history. Whales, birds or fishes may migrate thousands of miles throughout their life. In part, this is an adaptation process that the species has evolved during changing environmental conditions to ensure that the niches they occupy during breeding or feeding are optimal. However, this is a powerful selective phenomenon and many individuals die as a consequence of migration.

How the long migratory journeys of some animals evolved is not clear. It may have begun initially as a shorter journey to gain access to preferred foods or breeding conditions. Most animals have a territorial range they traverse throughout their life to seek these conditions but these migratory patterns change as a result of the animal's adaptation to geological and climactic conditions on the planet. Over the course of millions of years, environments change because of continental drift, polar ice cap contraction and expansion, formation of new mountain ranges, valleys and lakes, for example. Thus, the migration may become longer and more convoluted as these optimal foraging and breeding environments become separated.

The physiological needs of an animal surviving over a journey of thousands of kilometres are much greater than those for an animal needing only to migrate ten kilometres. Individuals that do not have the capacity to fulfil those needs may die during these migrations. Ultimately, this is related to the animal's energy reserves, which clearly relate to the amount of nutrition that the animal can take in before engaging in the migration. A number of situations can occur that will interfere with an animal's ability to gain enough nutrients. This could involve lack of access to food due to changing conditions, increased predators and disease, for example. Fewer nutrients will curtail the ability to migrate. In general, there is little foraging activity during migration allowing only supplementary increases in nutrition. Insufficient energy will reduce the animal's ability to ward off predators, pathogens or deal with the stress of excessive locomotion on a daily basis. For example, whales that are washed up on beaches during their migration sometimes show signs of malnutrition.

The major niche shifts resulting from migration are frequently associated with metamorphic or other transformative processes that impact on a variety of physiological systems. The interactions of these physiological processes during migration affect survival, growth and, ultimately, reproductive success. Animals that experience significant changes in their environmental conditions that affect fitness as part of their normal life history may gain a selective advantage by altering their capacity to survive under those conditions. But even animals that do not migrate may experience alterations in environmental conditions that affect survival and reproduction. There are a number of examples of seasonal changes in the capacity of animals to digest food or to withstand temperature extremes. Some studies in fish, for example, show that prior exposure to low temperatures improves swimming performance and growth at those temperatures. Taken together, the effect of migration per se can have a number of positive effects on fitness by increasing the environmental experience that a population is subjected to. Thus, individuals are selected, in part, on their ability to withstand these changing environmental stressors.

Glucocorticoids have been implicated in both the ability of animals to adapt to these stressors and, paradoxically, their death during migration. For example, cortisol can increase the incidence of downstream migratory behaviour in masu salmon. Similarly, activation of CRF and glucocorticoids are also implicated in the metamorphosis of tadpoles to adults in the western spadefoot frog. These species begin their metamorphic process as the ponds begin to dry up, and so the ability to change into adults becomes a mechanism to survive the loss of their previous habitat. The inter-regulation of stress and reproduction has been particularly well explored in migrating salmon. Plasma cortisol rises dramatically during the spawning migration of a number of salmonids. In fact, cortisol overproduction has been implicated in the post-spawning demise of these species. As we described in Chapter 3, high cortisol levels can inhibit feeding. Thus, in many cases, the death of post-spawning salmon may occur, in part, through malnutrition. Interestingly, these studies also indicated that gonadotrophin-releasing hormone (GnRH) production during reproductive readiness is not the cue for cortisol production. The system appears to have evolved to support spawning but is not triggered by it.

5.5   Reproductive strategy and habitat erosion

The human population of the planet has increased dramatically in recent decades, bringing with it a huge increase in anthropogenic activities that has led to a loss in wildlife habitat. The close approximation of human civilization and the resulting loss of habitat have increased the pressure on the reproductive capacity of animals through loss of food and increased predation and has also changed the structure of ecosystems and modified predator–prey interactions. This causes a level of stress on wild populations they did not have to deal with in the past. As many populations and species teeter on the edge of extinction it has been suggested that we could save some species by keeping them in managed herds and in zoos. These represent a novel and sometimes artificial environment in which the animals would need to adapt to new stressors. Thus, we need to understand the mechanisms of the stressors in order to manage these populations and species. It also creates a stress bottleneck in the natural evolution of these species, leading to selective pressure on the gene pool, selecting for those who can survive the new living conditions. Ultimately this selective pressure may also affect the ability to tolerate stressors. Thus, future novel stressors may have a different or stronger impact on the reproductive capacity of these animals. The loss and fragmentation of habitat due to the encroachment of human civilization and activities could conceivably affect the reproductive strategy of species.

Despite its success, sexual reproduction is costly and has a number of inherent problems. It reduces the number of individuals in a population that can be involved in procreation, recombination of DNA can lead to the break up of favourable gene combinations, and there can be a considerable cost associated with mate search and courtship. Asexual reproduction has many advantages over sexual reproduction but has main disadvantages of the accumulation of deleterious mutations and the genetic uniformity in the offspring which would slow down the rate of evolution. Three types of unisexual reproduction have been recognized in vertebrates. Hybridogenesis occurs when meiosis has been altered so that only females transfer their genome to the oocyte. The male genome is lost. Examples of this are found in fishes and amphibians. A second unisexual strategy is termed gynogenesis. In this situation, the females produce diploid eggs but need sperm from a closely related sexual species to trigger the onset of embryonic development. The male genetic material is inactivated and does not contribute to the offspring. Thus, the offspring are genetically identical to the mother. A well-studied example of this strategy is the Amazon molly. Parthenogenesis, the third recognized strategy, is defined as the development of an embryo from a female gamete without any contribution from male gametes. This is found typically in reptiles.

Recently, the identification of unisexual reproduction in the Komodo dragon and the hammerhead shark in captivity has highlighted the concern as to whether certain species have the ability to switch between sexual and asexual modes of reproduction in the wild. In both species, the individuals had been separated from males. Thus one might imagine that with the loss and resultant fragmentation of habitat, the probability of finding a suitable mate may be decreased. This facultative parthenogenesis may be an evolutionary adaptation to allow for continued reproduction when mates are unavailable. Komodo dragons and other monitor lizards utilize the ZW/ZZ sex determination system (see Chapter 2), in which the female is the heterogametic sex. The male results from a ZZ phenotype. Thus, monitor parthenogens are all male. The hammerhead shark, on the other hand, uses the XX/XY system and, therefore, all parthenogenic offspring are female. Presumably the development of male parthenogenic progeny in monitor lizards could allow a breeding pair in the form of the progeny and the female, but this will induce considerable genetic homogeneity. In hammerhead sharks, the fact that parthenogens are exclusively female will skew the population toward females. This may allow the population to pick up quicker, but it will also suffer from the expected loss of fitness from genetic homogeneity. However, there is evidence to indicate that species capable of facultative parthenogenesis may recover quicker from a population decline as a result of a natural disaster or colonize new areas quicker than sexual species. Recall also from Chapter 2, that species may switch from XX/XY and ZZ/ZW sex determination. It is unclear how often this has occurred in the past, or what the mechanisms are to foster such a switch, but it seems likely that as habitat and niches become compromised and greater numbers of species are living in captivity, we would expect to see more of this phenomenon.

Currently, facultative parthenogenesis has been documented in most classes of vertebrates, including mammals, but the production of viable offspring has not been recorded in mammals. In the case of parthenogenic mice, for example, the embryos die early in development. This suggests that the mechanisms governing mammalian parthenogenesis may differ from those in physiologically less complex species. In mammals, under normal circumstances, genomic imprinting (see Chapter 7) ensures regular embryonic development; the parental genomes have an effect on the expression of genes in the progeny. In other words, in order to ensure the normal development of mammalian embryos, genes from both sexes are necessary. In evolutionary terms this need may relate to the behavioural complexity in mammals. Behavioural expression in mammals is generally much more complex than that in non-mammals. The complexity of the mammalian behavioural repertoire has provided mammals with the ability to adapt to niche and habitat variation by behavioural adaptation. Thus behavioural adaptation, as found in mammals, may have offset the need for some physiological and genetic flexibility and, as a result, some of the processes that allow viable progeny of parthenogenic individuals in other classes of vertebrates have been lost in mammals.

Currently, facultative parthenogenesis has been documented in most classes of vertebrates, including mammals, although mammals do not appear to be able to produce viable offspring. Under normal circumstances, genomic imprinting (see Chapter 7) ensures regular embryonic development and thus in the case of parthenogenic mice the embryos die early in development. The expression of a gene may be related to its parental origin. This situation occurs in which the parental genomes have an effect on the expression of genes in the progeny and is termed 'genomic imprinting'. The result of imprinting is that individuals bearing two copies of either maternal or paternal alleles are unbalanced in terms of gene expression. Behavioural adaptation, as found in mammals, may have offset the need for some physiological and genetic flexibility. The interaction of complex neurological and other physiological adaptations is poorly understood. A frequently cited reason is that the changes required in the development of the brain are too great to allow such adaptive changes.

Hermaphroditism is another strategy that could be employed if population numbers dwindle to a critical level. Hermaphroditism, where an individual is a functional male and female at some time during their lifespan, is most common in fishes and particularly coral reef fishes. The animal can be either a simultaneous hermaphrodite, when they are a functional male and female at the same time, or a sequential hermaphrodite where they are born as one sex then switch to another sex at a later period in their life. The latter is the more common strategy of the two.

5.6   Human industrial waste as an evolutionarily novel stressor

Sadly, a wide range of anthropogenic chemicals released into the environment have the capacity to disrupt the endocrine and nervous systems in metazoans. This leads to abnormalities in growth, development reproduction and behaviour. Pesticides that mimic oestrogens have been shown to alter sexual and reproductive behaviours in birds. In mammals, polychlorinated biphenyls (PCBs) and DDT-derived compounds interact with steroid hormone receptors, thyroid hormone receptors and the aryl hydrocarbon receptor. Nonylphenol can act as an oestrogen agonist or androgen antagonist and can alter gonado-trophin synthesis and secretion.

In a number of amphibians and ray-finned fish there is considerable plasticity in the reproductive development in adults. The presence of exoge-nous steroids or steroid-like mimics can have significant effects on behaviour and development. In some rivers, industrial waste has induced full or partial sex reversal in most of the male fish.

Nonylphenol decreases gonadotrophins and increases vitellogenin produc-tion in the livers of male rainbow trout. Naphthalene and β-naphthaflavone, on the other hand, increase steroidogenesis in testes. And in the killifish, bisphenol A and nonylphenol induce vitellogenin synthesis in the male liver. In guppies, DDE and the fungicide vinclozolin have been implicated in reduced male colouration, reduced spermatogenesis, induced altered courtship behav-iour and decreased testes size.

In recent years there has been a worldwide decline in amphibians. A number of the chemicals that mimic steroid, thyroid and possibly retinoic acid hormones can disrupt metamorphosis in frogs, leading to developmental abnormalities. Methroprene and atrazine have been identified as possible thyroid hormone disruptors. In some frogs, as little as 1 μmol/L (ppm) of dibutylphthalate induces the formation of ovaries in some of the males. Exposure to DDT and DDE can reduce male colouration and promote female colouration in males. Male frogs generally absorb more of the xenobiotic than female frogs. Also in frogs, bisphenol A or methoxychlor can increase vasotocin to produce antidiuretic effects and decrease the diuretic effect of mesotocin.

Many other compounds produced by human activities also have deleterious effects on the reproductive capacity of animals. Tributyl tin is a component of an antifouling agent that is applied to ships' hulls. It can accumulate in the ganglia of gastropods and has been implicated in the imposition of male characteristics on female molluscs. It appears to inhibit the actions of cytochrome P450-dependent aromatase, thus increasing the levels of andro-gens, and may also affect the actions of an neuroendocrine factor. Although the factor has not been conclusively determined, a neuropeptide called APGWamide has been shown to induce male-like sex organs in some female molluscs and it is thought that tributyl tin may be acting as a hormone mimic on the APGWamide receptor. A number of pesticides developed to control insects have been designed specifically to interfere with the endocrine system, for example, to disrupt the mechanisms of ecdysterone or juvenile hormone receptors. However, as we have seen in previous chapters, there is considerable functional similarity between the steroids of arthropods and those of vertebrates and so the effects of such chemicals may not be confined to their target animals.

5.7 Anticipation of stress

Species are under intense selective pressure to adapt to their environment. Adaptation consists of a multitude of factors. For the most part of this book we have been discussing an organism's adaptation to environment with respect to physical attributes such as oxygen availability, temperature, salinity, pH and food sources, for example. However, there is also a survival advantage if animals are temporally adapted to their environment. During the course of a 24-hour period, light intensity, temperature, humidity and composition of organismal interactions vary. Over the year, seasonal changes in the form of temperature, photoperiod and precipitation regulate food availability, predator–prey relationships and habitat availability. For many species, events such as migration, hibernation and reproduction are timed to particular periods in the season and require significant physiological preparations.

Thus, the ability to anticipate changes in the environment provides a selective advantage over species that do not have this ability. Physical or temporal adaptation maximizes the efficiency through which animals maintain homeostatic mechanisms. Homeostasis may be classified as being reactive or predictive. Reactive homeostasis occurs in response to a homeostasis-challenging event, whereas predictive homeostasis is anticipatory, for example when an animal forages for food and anticipates when a food source will become available.

Metazoans have evolved internal clocks in the form of molecular and physiological rhythms that allow the anticipation of temporal changes in the environment. The endogenous biological rhythms may be synchronized with one or more naturally occurring cycles, such as the periodicity defined by diurnal, lunar, tidal or seasonal cycles. Three basic types of periodicity are recognized. Circadian rhythms are those that are about a day in duration. Ultradian rhythms occur with a periodicity shorter than a day. These periods may be in the order of minutes to hours. Many of the pulse times of hypothalamic-releasing factors, pituitary hormones and the peripheral hormone systems they drive fall into this category. However, a number of species that follow tidal rhythms also fall into this category. Finally, rhythms with a periodicity of greater than a day are referred to as infradian rhythms. Examples of infradian rhythms include seasonal, annual and oestrous or menstrual cycles. For an apparent cycle to be classed as an endogenous rhythm it must be shown to persist in the absence of cues associated with that rhythm. Moreover, it must be shown to have the ability to be entrained by environmental cues. We might think of the entrainment of rhythms as the equivalent of resetting a watch periodically so that it remains in harmony with other clocks.

Seasonal fluctuations in immune function have been well documented, for example the number of circulating leukocytes in mice, rats, voles and humans are elevated during autumn and winter. Also, spleen and thymus masses change in deer mice and prairie voles during autumn and winter. These changes are mediated by day length. Short day exposure increases splenic masses in deer mice and the Syrian hamster, as well as elevating total lymphocyte and macrophage counts.

5.8 Nutrition, toxins and infertility

Optimal reproductive performance is achieved when the diet is balanced and meets specific requirements for growth maintenance, pregnancy or lactation. Specific nutrients may be required for some of these functions. For example, a lack of vitamin A can affect spermatogenesis. Restricted nutrition probably delays puberty via inhibition of the pulsatile release of GnRH. In women, amenorrhoea occurs when the body weight falls to about 75% of the ideal body weight. Improved nutrients can increase the ovulatory rate in some animals. Although it is not clear how this happens, insulin can increase ovulation rate. Granulosa cells have receptors for insulin-like growth factor (IGF)-I and growth hormone. Growth hormone also plays a synergistic role with the gonadotrophins and sex steroids for ovarian and testicular function. An initiation of reproductive activity which follows increased feeding of growth-restricted lambs may be due to the increased availability of tyrosine. However, a high level of feeding in early pregnancy can also be detrimental possibly due to a decrease in progesterone caused by the stimulatory effect of high food intake on hepatic blood flow and progesterone metabolic clearance rate. In menopause women experience a period of suboptimal ovarian function before the oocytes are expended.

For conception, pregnancy and birth of live young to occur, an ordered sequence of events must happen. This includes sperm production, maturation and transport of sperm, effective transfer of sperm to female, synchronization of sperm disposition with ovulation, further sperm maturation and transport in the female, fertilization, embryonic maturation and transport, endometrial development for implantation, maternal recognition of pregnancy and corpus luteum maintenance, embryo and fetal development and finally successful parturition. Any of these events can fail, causing a reduction in reproductive viability.

5.8.1   Nutrients and toxins

A number of heavy metals when ingested in excess are toxic to both sexes. However, because many heavy metals are required for certain biochemical reactions, they are also nutrients in low concentrations (Figure 5.2). Thus, either the loss of such metal ions through aberrant diets or exposure to toxic levels can affect reproductive function.

Spermatogenic failure has been associated with disease and injury induced by heavy metals and toxins. Vitamin E and B deficiency may affect fertility in male mammals. Vitamin D deficiency can retard spermatogenesis by interfering with the function of Sertoli and Leydig cells. Reduced food intake affects Leydig cells more severely than seminiferous tubules. Some chemicals, such as phthalate acid esters, can cause a decrease in zinc. Numerous heavy metals are found in normal human semen as a result of nutrient requirements and toxic exposure. Chemicals and heavy metals may induce mutation in the male germ line. Acute cadmium exposure induces vascular destruction in the testes. This hypoxic ischaemia produces a number of secondary effects including seminiferous tubule necrosis and Leydig cell damage. The permeability of the seminiferous tubules is increased, leading to increased toxicity to other chemicals. Sertoli cells are targets of phthalates and other chemical and xenobiotics. Cadmium, in particular, has been implicated in testicular pathology. Cadmium may interfere with the cellular messenger cAMP in Leydig cells,

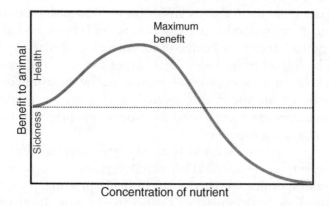

Figure 5.2   Relationship between nutrition and toxicity. All nutrients have an optimum intake that will benefit the animal. If the intake is too high, then it will act like a toxin. Sufficiently low concentrations of any toxin can also act like a nutrient. Note the similarity with Figure 1.2

thus disrupting the actions on these cells of luteinizing hormone and other hormones that require this messenger. Leydig cell production of testosterone is reduced by cadmium and other metal cations. ROS production can also lead to gonadal disruption.

Zinc uptake by the male reproductive system is androgen-dependent and is required for normal function. There are many zinc-containing metalloenzymes and transcription factors. Thus, a reduction of dietary zinc can inhibit a number of the normal physiological processes in reproduction. Zinc toxicity is not common in animals, including humans, but zinc deficiencies are. Zinc deficiency can occur via poor diets or through xenochemicals that interfere with normal zinc metabolism. It can lead to a severe atrophy of the germinal epithelium, resulting in reduced sperm production and, in young males, can inhibit normal sexual development. This effect can sometimes be reversed by giving zinc adequate diets. Testes and epididymis contain a high activity of angiotensin-converting enzyme (ACE) which is also a zinc metalloenzyme. ACE activity is found primarily in germ cells.

Lead toxicity has been associated with infertility, germinal epithelium damage, decreased sperm motility, decreased testosterone concentrations, abnormal sperm and interference with androgen receptor binding. Lead can accumulate in many regions of the testes. Lead has also been linked to a number of neurological effects that can lead to altered behaviours and death.

Heavy metals have been better studied in the testes due to the relative simplicity of the system. In general, the pathways are complex and involve a number of physiological mechanisms. In females, lead poisoning has been associated with reduced fertility miscarriages and stillbirths. Cobalt may interfere with DNA repair mechanisms. Cobalt toxicity may be also related to the general hypoxia that occurs after exposure to the metal. In females, high mercury levels can prolong the activity of the corpora lutea, causing oestradiol and progesterone to remain elevated in non-pregnant females and resulting in changes in oestrous or menstrual cycles. Cadmium has been associated with necrosis of follicles and extensive haemorrhage, whereas lead can inhibit steroidogenesis. Zinc deficiencies may induce high levels of prolactin and inhibition of the hypothalamic-pituitary-ovarian axis.

5.9   Summary

Stress is a consequence of living. Environmental stressors may be classified as biotic – originating from other living species – or abiotic – resulting from climatic or geological disturbances. Changes in the environment happen all the time and animals have needed to evolve a number of mechanisms to anticipate and protect themselves from stressors resulting from environmental change. Each type of organism, whether classified as poikilothermic or endothermic, will have its own unique set of challenges. Organisms have evolved a number of sexual and asexual reproductive strategies to contend with changing environmental conditions and habitat loss. In addition, all organisms live at the cost of other organisms and a number of defence mechanisms have been evolved. As a result of human civilization, numerous toxic chemicals have been released into the environment. Exposure to many of these chemicals can induce damage to reproductive tissues, leading to a loss of reproductive ability.

6

Saving women and children first: protecting the progeny

> ...the real importance of a large number of eggs or seed is to make up for much destruction at some period of life and this period in the great majority of cases in an early one
>
> Charles Darwin, *On the Origin of Species* (1859)

6.1 Introduction

During the sinking of the *Titanic*, when the hopelessness of the condition became apparent to Captain Smith, he ordered his officers to load the women and children into the lifeboats. No men were allowed into boats until all the women and children were loaded first. Of course this didn't work out exactly as planned as the need to survive is a very strong one. But this approach reflected a long-standing practice in maritime custom. And it is generally believed that instinct drives us to save our children and sacrifice ourselves so that our children may live. This isn't always the case of course. Usually, the greater the investment we put into the raising of our progeny, the greater the need to protect them.

Evolution of the reproductive system developed to ensure safety of the progeny in a stress-free environment. There is a large investment in progeny in all species. This investment differs with respect to physiological, in utero and behavioural effects and varies considerably between males and females. Both sexes engage in activities to ensure that their genes are passed on but they express it in very different ways. Males tend to put energy into defending

Sex, Stress and Reproductive Success, First Edition. David A. Lovejoy and Dalia Barsyte.
© 2011 John Wiley & Sons, Ltd. Published 2011 by John Wiley & Sons, Ltd.

the female from predators or other males, whereas the female tends to put the greatest energy into protecting their progeny directly. This, then, puts a different type of stress on males and females.

In his book, *On the Origin of Species*, Charles Darwin recognized that there is not a great correlation between the number of eggs laid and the number of viable progeny. He further recognized the importance of parental investment in the care of the progeny. He also pointed out that the amount of food that each species represents provides a limit to which other species can increase. However, frequently it is not the obtaining of food, but the extent that each species serves as prey to other animals that determines the average numbers of species. Ecosystems balance out so that an animal may be selected, in part, for the large number of progeny, so that they will feed other prey to allow the preservation of a few individuals.

6.2 Sexual selection costs and stress

The selection pressures that males and females exert on conspecifics as they acquire mates produce a category of natural selection called sexual selection. Sexual selection occurs because there is a correlation between the gender of an individual and its parental investment for each offspring. Parental investment is whatever increases the probability that some offspring will survive to reproduce at the cost of the parent's ability to generate additional offspring. Males and females interact in sexual and reproductive manners in the form of courtship, copulation and care of offspring. This could be thought of as a single common cause uniting two inherently selfish partners. Both male and female reproductive physiology has evolved to a certain extent independently in each sex to maximize their reproductive potential (Figure 6.1).

Although a female is defined as the sex that produces the ova, she may also be defined as an individual that makes a relatively large parental investment per gamete. For example, in many birds, the weight of a single egg can be 15–20% of the weight of the female, and in some cases may be as high as 30%. For many species, a clutch of eggs may be heavier than the female. Males could be then defined as individuals whose gametes contain genetic information and little else. Yes, this is a little harsh, but in the cold objective world of science, this is frequently what it comes down to. The amount of sperm produced in birds is usually only a small percentage of body weight and frequently far less. But these numbers can be misleading, because males are generally producing vast quantities of sperm throughout their life.

In mammals, the ejaculate in copulations is energetically more expensive than single egg production, but the energy demands associated directly

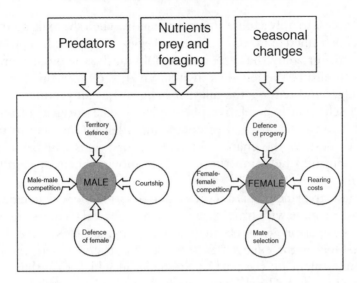

Figure 6.1 Key stressful interactions during courtship, mate selection and rearing of the progeny

with fertilization and development, which falls on a female mammal after copulation, are vastly greater than those imposed on the male. Having said that, it should be noted that a number of males in different species expend a large amount of energy defending the nest, territory and female from predators and other males during development.

The individuals of both sexes are expected to behave in a way that maximizes the number of genes passed to the next generation. In most situations, females engage in far greater parental care than males. Parental care is a form of parental investment that raises the survival chances of an already existing offspring but decreases the probability of producing additional offspring. In other words, it is better to guard my treasure now, rather than throw it away in the hopes that I might acquire a bigger treasure sometime in the future. Given that the energetic cost to a male of a single copulation is far less than that to a female, including post-copulatory actions, then the female has far more to lose. A mistake is far costlier to her. Moreover, whereas a female can guarantee she has investment of at least 50% of her genes in the progeny, the male has no such guarantee. Perhaps for these reasons, females may be more sensitive to the effects of stress than males with respect to reproductive function. At very little physiological cost, a male can transfer genes to offspring that will receive physiologically costly support from their mother.

Having access to only a finite amount of energy when she makes gametes, the female is constrained in the total number of eggs she can produce, which limits the number of progeny she can produce. Obviously, the more energy that is expended to ensure the female's own survival, the more this cuts into this energy budget and reduces the number of eggs she can produce.

Natural selection is a relative phenomenon. The survival of an allele is measured against the survival of competing alleles. One way to maximize fitness is to destroy the chances of competitors. This can be achieved by killing or injuring the competitors or by removing their opportunities for copulations.

The 'handicap principal' is a term used to describe the way females of some species prefer a male with traits that reduce his survival chances but announce that he has superior genes because he has managed to survive despite his handicap. One might expect this to occur in species where there is little parental investment on the part of the male, in which case they mate then are no longer needed. The elaborate feathers of a bird of paradise or a peacock have the cost of lowering the survival chances of the males but have the benefit of announcing the competence of the surviving male.

In some species a male will give up opportunities to copulate with other females to assist the investment of his progeny. In general, parental care becomes advantageous when food resources are difficult for young animals to secure, either because the prey is difficult to capture, or there is strong competition for these resources, or when the young face predators that the adults can repel but not the young. Maternal care may become enhanced in situations where there is a lack of reproductive extremes. In such cases, the male's production of sperm does not have an advantage for the male. In some cases, there may be a protracted courtship, as is the case with some gulls, such that by the time the courtship is over, there are few receptive females remaining. Thus, the best use of the males is to invest in his progeny. When a male can provide important resources or services directly or indirectly to his offspring a female may select a partner whose behaviour, appearance or territory indicates that he will contribute more to her fitness than other males. During a particularly long courtship, a female can assess the male's willingness to provide parental investment.

In many species, the presence of a novel male that successfully removes the mating male will frequently destroy the young, for example in lions and mice. Thus, it is in the female's best interest to select a male that can aid in the survival of the young. Typically, males compete to copulate but as males provide an increasingly large parental investment per offspring, females begin to compete more and more among themselves for opportunities to mate and males become more selective about their partners.

6.3   Male–male interaction stressors

A male's fitness tends to be proportional to the number of females he fertilizes in his lifetime. As a result, this generates intense competition among males to gain access to the female. If a male impregnates a female and the progeny is not viable, it is at little cost to the male. For a female, on the other hand, the cost is much greater. For this reason, females are likely to be more stringent about selecting an appropriate mate than males. Males are aware of this and frequently go to extreme lengths to gain access to the female. A male's reproductive capacity is based on a number of aspects. In order to improve fitness, he can improve the opportunity to copulate with receptive females, or he can increase the probability that the sperm will be used to fertilize the eggs. Alternatively, he could reduce the fitness of competitor males.

So, how do males increase their opportunity to mate with receptive females? Males tend to have a lower threshold for sexual arousal. It has been thought that because of the low cost associated with the male's copulation, the opportunity to mate, when it arrives, is advantageous to the male and, therefore, this trait has been selected for. This rapid arousal of males increase the chances that they will not miss any opportunity to pass their genes on, but also means that males are more likely than females to engage in biologically relevant sexual behaviour.

The importance of copulation is seen in the willingness of males to fight for females and the evolution of structural adaptations such as horns or claws. We might wonder, however, how much of the selection pressure of the development of these modifications is due to inter-male rivalry and how much is due to predation pressure. It is likely, of course, that, depending on the species and the ecosystem in which they live, these traits have co-evolved. For example, inter-sibling rivalry – for example in the form of lion cubs playing – has a role in developing defence behaviours, a critical component to the defence from predators.

Now, what happens when you introduce increased levels of stressors into the life of males? Let's consider an average male vertebrate possessing the traits described above. We might expect that an appropriate level of stress will act on all three of these sexual prerogatives. Aggression is expensive both in terms of calories and associated risk of injury. Assuming that stress shunts the energy balance to defence and away from reproduction, we would expect to see a decrease in sexual- and courtship-related aggression in response to stress. This raises the question as to how much male aggression can be attributed to courtship. The amount of time spent warding off other males is stressful.

Animals have evolved a number of different strategies to ensure that certain males have access to females. We refer to these males with preferred access to females as 'dominant males'. In the most basic sense, a male might simply 'bully' and antagonize other males to keep them away. Alternatively, a dominant male may travel with harems of females. In zebras, for example, the male will rigorously defend his harem from all other marauding males. The advantage of this approach is that because the females are not all receptive at the same time, the male can copulate with each one of them as they become available. If a male's aggression tactics make a lasting impression on other males in its social group, then this may give rise to dominance hierarchies. This aspect of reproductive behaviour is widespread in vertebrates and arguably well-entrenched in human behaviour. In dominance hierarchies, the hierarchy may be established initially through aggressive interaction among individuals. Once this hierarchy is established, then the dominant male may endure for some time without subsequence challenges. The advantage of dominance hierarchies is that it reduces the stress of repeated challenges and can allow the dominant male to be more successful at passing on his genes. The dominant individuals have preferred access to a number of resources, such as better resting spots and superior food resources, as well as females. Fighting is costly, but for many species the dominance reign is brief. For a number of species with hierarchies, for example elephant seals, the dominant males have a shorter lifespan than the females they defend.

Dominance interactions are also the basis for territorial behaviours. Here, the male defends an area rather than a female against other males. The males complete for the territories that have the best resources. If a male can claim a good territory, he will have access to the females that come to him. With superior resources available to rear the young and provide protection, he increases the number of females that will choose his territory. Naturally, there are exceptions to this. For the American jacana, for example, it is the female bird that defends the territory and will mate with all the males who tend the nests. In most species, however, it is the male that defends the territory. Territories may be trees, nests or ranges.

Non-territorial individuals also have an opportunity to breed. One of the problems of extreme dominance is that it limits the gene pool to those individuals who have been successful in defending against other males. This might mean that only the biggest and more experienced males are successful. And so over a number of successful generations, we might expect this trait to grow out of proportion with other traits. But environmental conditions are constantly changing between generations and a loss of adaptive ability is a loss of fitness. Depending on the species, however, there are a number of ways in which the subdominant males avoid the inter-male stressors entirely. There is a cost to dominance encounters. Long drawn out bouts, if continued, will eventually wear down the dominant male over the course of multiple challenges. Therefore, it is in the best interests of both the challenger and the dominant male to resolve these bouts as quickly as possible. Perhaps for these reasons, ornamentation in males signifying dominance is a way of signalling to potential challengers not to bother with a challenge because they will lose. In addition, there may be a number of non-contact behaviours that act to settle these challenges. Thus in order to maximize fitness, a subordinate should conserve his energy resources and wait for an appropriate time in the future when he might have an advantage. In other words: 'he who fights and runs away, lives to fight another day'.

Another approach to gain access to the female is to copulate with her while the dominant male is not present. I mean, he can't be everywhere at once, can he? Males employing this strategy are given the less than flattering name of 'sneaker males'. As far as evolution is concerned this is a strategy that works. These males have been selected for traits other than those that will lead to a head-to-head combat with the dominant individual but still will allow copulation with the females. The dominant individual is busy defending his females, his status and his resources.

Female mimicry is another strategy used by males of some species. In such a strategy, the subdominant male has traits that are more similar to those of a female than the dominant male. This occurs most commonly in fish although it has been documented in other classes of vertebrates. It may have evolved from the effect of some stress, perhaps inter-male aggression. In such a scenario, the activated stress response inhibits the reproductive system sufficiently that there is a reduction in secondary sex characteristics, for example colour, or perhaps pheromones. The affected individual traits take on a more female-like appearance and, therefore, when he approaches the dominant male, the subdominant male is not recognized as such by the dominant male and so is ignored. The subdominant male then gains access to the female and fertilizes her.

6.3.1   Courtship costs

Courtship costs differ considerably between the sexes. In the past, courtship rituals have been investigated in terms of synchrony of sexual arousal and the forming and developing of pair-bonding. Over the last few decades, with our obsession about the mechanistic approaches to life as defined by our molecular biology revolution and the enormous influence of Richard Dawkins' 'Selfish Gene' theory has altered our philosophical approach to the study of courtship. Thus, these older concepts are being replaced by new views that focus on the conflict of interest between the sexes, or 'the battle of sexes' and the 'sexual arms race'. This change from views of harmony between the sexes to one of antagonism between the sexes over the last sixty or so years is interesting and probably reflects changing conditions within our own human society.

Factors that reduce a parent's later reproductive output include the debilitating effects of meeting the energetic demands of reproduction, as well as the heightened risk of predation incurred while attempting to produce offspring. The cost of the progeny is directly related to the costs of the maturation, courtship and reproductive costs of the parents. Few environments are uniform. They vary between location seasons and over the years. Because of this there is no guarantee even if an individual offspring were to live and reproduce themselves in exactly the same spot where the parents lived and reproduced that the environment would be the same, or even necessarily favourable to the development of the young. In other words, the condition of the environment that the progeny are to be reared in cannot be anticipated.

In sexual reproduction, each individual is essentially unique and will necessarily differ from their parents in a number of respects. This places a limit on the number of complex behaviours that can be genetically encoded and, therefore, a certain amount of learning is necessary. It does not seem possible for a sufficient number of fixed behavioural patterns to be encoded genetically that will allow the organism to anticipate each new stressor that arrives. It is not possible to be selected for one trait on the off-chance it becomes useful several generations in the future.

6.3.2   Mate selection

When I was a teenager and started dating, I was convinced that I needed a car to attract girls. A considerable amount of attention was devoted to this, and I, like many of the other boys, debated the merits of one car over another as to which was the most efficient at attracting the girls. The sexiest cars were the European sports cars, but none of us could afford those. What was affordable in the small town where I lived were the muscle cars. These were the 'souped-up' American cars with big engines and tyres. The muscle cars represented strength and masculine vitality. Ironically, you could buy these relatively cheaply, then spend the rest of your money on improving it. We all instinctively understood, in our adolescent logic, that girls made the decision as to who they were going to date. Thus, in order to attract a girl's attention we needed a large 'macho' car, like male birds needed large colourful feathers, or reindeer needed impressive antlers. Somehow, that seemed easier than trying to act like humans and actually talk to girls. This was particularly stressful to me, and I did not try to date until I had a car.

Traditionally, mate selection was based on the concept that the female chooses the mate best suited to survive in a hostile environment on the basis of physical attributes designed to indicate that the male is better suited to handle challenge and adversity. Examples of this might include larger body size or larger antlers, or perhaps greater aggression and strength as evidenced by males winning bouts over other males. Such concepts can be traced to Darwin's writings although he based some of his ideas on sexual selection and mate competition on the social mores current in Victorian England at the time. However, Darwin also observed that many females select their mates on the basis of traits that had seemingly little to do with a male's ability to win against competitors. Examples of this are the colourful facial marking of male mandrills, the tail feathers of peacocks, the croaking of male frogs or the singing of male songbirds.

Later, these observations were explained by the possible correlation of these traits with hormones and other physical attributes that drew the attention of the female to the male. For example, a larger antler size or a greater size or colour of feather indicated to the female that this was a more robust male that could survive and presumably pass on these traits to the progeny.

However, in some cases, the large tail feathers of a bird of paradise, for example, can impede the cock's flight and may also attract predators. This led to additional theories, in the case of birds, to suggest that these colourful markings were designed to do just that, to attract predators and lure them away from the female. In the case of the bird of paradise tail feathers, the extra weight and bulkiness may act indicate to the predator that it is more likely to catch the male causing it to prefer to hunt him rather than the female. To use another example, the red-winged blackbird typically stays near the female during the courtship and brooding season, but will take flight to a nearby bush at the appearance of a potential predator. The red and yellow markings on the front of the wing are particularly evident during flight and more likely to attract a predator. The female, on the other hand, is well camouflaged with cryptic colouration and is not likely to have been seen. The red-winged blackbird male is particularly aggressive, and will defend the nest and female from other males and predators. Initially it was thought that is was an example of male parental behaviour in that this behaviour helps ensure the male's genes are protected. However, studies have since indicated that in many cases other males gain access to the female while the defending male is away. Moreover, aggressive defence of a nest can lead to physical exhaustion of the male with the result that another male mates with that particular female. Here, we might expect that the strongest and most aggressive males might be more successful in passing on their genes, and less robust males less successful in this regard. However, as we pointed out earlier, some 'sneaker' males, although perhaps less robust, may be particularly attuned to the activity of a dominant male, and successfully mate with the female while the male is away.

If one thinks of this in the context of the proliferation of certain genes, we might consider then that there is a certain competition among genes to survive. Thus, the set of genes associated with one trait, for example antler size, may initially produce male offspring that are more able to defend themselves, thereby increasing the probability that they will mate and therefore allowing their genes to survive. If this is allowed to continue and the genes lead to increasingly large antlers, the individuals need to compensate with greater physical strength and hence greater energy demands. If this is allowed to continue, then the weight and mass of the antlers may start to decrease the ability of the individual to adequately defend itself. Should a more lithesome and energetic male with smaller antlers appear, it may win the next bout. In this case, the genes regulating the attributes of physical traits designed to defend against competitive stress eventually lose out to a set of genes that are associated with energy production which are ultimately associated with the stress of sufficient nutrition and foraging.

We might imagine this as an ecosystem of genes and their protein products, many of which are in competition with each other. But how successful these genes are is dependent, in part, upon the external ecosystem. For example, in an environment that has few predators and is filled with high nutrition food sources, then antler growth may be accelerated in the population by the selection of these genes since the animals' chief competition is between individuals of the same population. However over time, with the arrival of other species that compete for the same food source, or changes in climatic conditions that impact the high energy food sources, then antler size becomes less important for the individual to survive. Another example of this might be female mimicry where these males engage in less male-to-male competition but have greater access to the females. In this manner their genes are passed on and perhaps this is one of the reasons these genes continue to survive in many species. The need to mate is strong and thus any trait that allows a competitive advantage is likely to be passed on.

6.4   Summary

Successful mating does not necessarily ensure that the progeny will survive. As a result, there are a number of strategies that have evolved to maximize the chances that progeny will survive so that they too can reproduce. One such strategy is to produce large numbers of progeny, to effectively feed the predators yet allow a few of the progeny to survive. Another method is to provide greater care and protection to the progeny by the adults, so that they can ward off predators. The cost of reproduction, courtship and progeny investment differs greatly between the sexes. The energy that either sex invests into these processes will reflect both the cost of the investment and the probability that their genes will be passed on. For most females, their genetic investment is guaranteed, but for males this may not be the case. Thus females tend to put most of their energy into the direct protection of their progeny, whereas males will put greater emphasis on protecting their access to females and their progeny by competing with other males and warding off potential predators. Depending up the type of strategy utilized, the sexes will be subject to stressors associated with that strategy.

7

Epigenetic factors in reproductive success: don't ignore your parents

For what is reproduction, but the building up of a new organism with a detached fragment of the old?

Henri Bergson, *The Evolution of Life* (1907)

7.1 Introduction

We generally think of genetic inheritances as the transmission of traits across generations, which, in the past, have been attributed by the inheritance of genomic information from parental generations. However, when we see that the same genetic material produces a slightly different outcome we wonder why. When I was in high school, there were identical twin boys in my class – always fun for the kids and the teachers. So many plays, films and books are premised on twins playing each other and being indistinguishable. In this case though, after one week there was no problem telling their faces apart, and the most amazing thing was how different those two guys were. These two might as well have been neighbours given the differences in their life. One fellow became a career-oriented artist and the other family-centred businessman. It was the same genetic information but with a different outcome.

There are millions of cells in the human body, from small round white blood cells to huge branched neurons that can reach from your brain down to the spinal cord. They all have the same DNA, same genetic material. All of the various shapes, sizes and function of the cells are the result of using that same DNA in a different way.

Epigenetics is the science associated with environmental (i.e. external to the organisms) and developmental effects on the mediation of the expression of

Sex, Stress and Reproductive Success, First Edition. David A. Lovejoy and Dalia Barsyte.
© 2011 John Wiley & Sons, Ltd. Published 2011 by John Wiley & Sons, Ltd.

genetically inherited traits. At a molecular level, epigenetics effects are the expressive capacity of a gene. Although sometimes we think of genes as being turned on or off, in reality they are expressed in gradients. Many genes may be partially expressed or partially inhibited, for example, the kitchen tap can be fully on or off or somewhere in the middle and sometimes leaky. The level at which the genes are expressed is dependent on the expression of the various transcription factors that regulate them. These transcription factors are expressed as a result of other hormones, chemicals and metabolic products they come into contact with. One way of regulating the DNA is through the epigenetic modifications of DNA.

7.1.1   Epigenetics and DNA methylation

The goal of seeking an understanding of how environmental conditions can induce long-term changes in DNA has a long history. In the past, it has been suggested that a long-term and stable association between DNA and proteins, histone acetylation, RNA interference or perhaps a chemical modification of the nucleotide bases in DNA could be responsible. Of these concepts, an investigation of the latter mechanism with respect to reproduction and stress has yielded a number of interesting findings that explain many of the observations seen in the epigenetic regulation of physiology and behaviour. Specifically, a process known as 'methylation' appears to play a major role.

Methylation is the enzymatic addition of a methyl group ($-CH_3$) to a nucleotide base. It is mainly cytosine bases that are methylated in DNA. The methylation reaction is generally confined to certain regions comprising distinct sequences in the DNA. The DNA methylases, the enzymes responsible for this reaction, use the chemical adenosylmethionine as the methyl group donor. In eukaryotic cells, about 5% of the cytosine bases in DNA are methylated to 5-methylcytidine. This is most common in cytosine–phosphate–guanosine (CpG) sequences, where cytosine is adjacent to the guanosine, and produces the modification symmetrically on both the sense and antisense strands of the DNA. These CpG sequences are not as common in the DNA as one would expect by chance as they are reserved for special regions of the genome. The sites where cytosine is methylated are generally found in the regulatory elements of the gene – segments that direct the expression of the gene but are not part of the transcribed product. By analogy, the head office of a car manufacturing plant produces no cars, but rather instructs the factories what to do and how much to produce.

In the cell nucleus, DNA is wrapped around a set of histone proteins (Figure 7.1). These histone–DNA clusters are known as chromatin. When

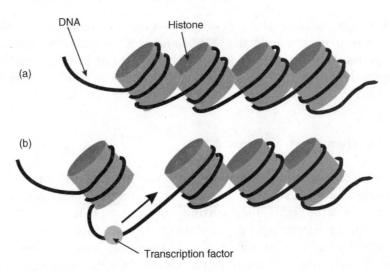

Figure 7.1   Interaction of DNA and histones. (a) Inactive DNA. (b) Transcriptionally active DNA

DNA is expressed, it must come into contact with the RNA polymerase and required transcription factors, and therefore must dissociate itself from the histones. After this dissociation, the DNA becomes active and the nucleotide sequences become exposed and available for the RNA polymerases and transcription factors to bind to. Transcription factors, as well as the polymerases, bind to particular nucleotide sequences. If the bases in the sequence are modified by, for example, the presence of a new methyl group, then these proteins cannot bind as well to this sequence. Thus, the functional effect of methylation is to inhibit the ability of the transcription factors to bind to the DNA, thereby reducing the expression of that particular gene. The methylation reaction is remarkably stable and can endure for a number of generations. It is not permanent, however, and can be reversed in some situations (Figure 7.2).

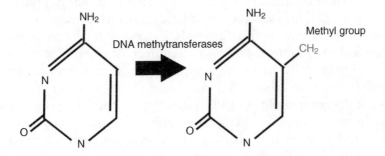

Figure 7.2   Methylation of cytosine

DNA methylation of the CpG sites is used by mammals to regulate transcription of genes, influence X chromosome inactivation, control imprinted genes and repress parasitic DNAs. Alterations in DNA methylation have been shown to lead to number of pathological conditions including cancer, male infertility and autism, to name a few examples.

7.1.2   Maternal effects of stress

Maternal effects on offspring have been shown in numerous species. As we have discussed previously, the longer the association of the progeny with the mother or parent the greater the probability of an effect. In mammals, during in utero development the offspring are particularly susceptible to hormonal and physiological effects from the mother. Depending on the physiological, nutritional and emotional states of the mother, a number of signals in the form of hormones and metabolic byproducts will be taken up by the embryo. Because the embryo is undergoing a profound set of developmental changes, these maternal effects have long-lasting effects on the fetus and may, in fact, remain for the lifespan of the offspring. We might think of these early developmental processes as laying down the basic foundation of the organism. The structure and development of this foundation will affect the development and manifestation of future processes.

Clearly, however, the length of time that the embryo, fetus or offspring is associated with the mother will influence the chances that the progeny will be affected by the physiological state of the mother. This would be expected to occur across taxa. Theoretically, the simplest organisms that reproduce asexually by, say fission or budding, could also be affected by the condition of the parent. We would expect, however, that more complex organisms would generate a greater variety of hormonal responses and associated metabolic byproducts.

We might also expect that the more neurologically complex an organism is, the greater the opportunity and ability to perceive changes in the environment. Again, if we were to consider a very simple organism that has only a simple sensory system, say, for example, an olfactory (chemical detection) sense, then its perception of the environment around it is dependent only on the perception of its chemical environment. This limits the effect the environment can have on the parent. On the other hand, consider an organism with a more complex set of olfactory, gustatory, touch and visual senses, not to mention other senses such as echolocation, audition and electroception, then the perception of the world is far more complex. This sensory input offers numerous opportunities to detect and process disturbances in the environment. Coupled with a memory of dangers of such disturbances, then the parent is far more likely to undergo a stress response because they can perceive the world in a much higher resolution. This then would further suggest that the more complex the neural processing the greater the opportunity to understand the significance of these environmental disturbances.

Putting this all together, we might expect that neurologically complex animals that are capable of numerous sensory modality perceptions and that possess a long gestation and rearing periods will be more sensitive and therefore affected by the condition of the mother and parents. Thus we have arrived at – you guessed it – mammals and particularly humans. Humans are probably the most sensitive of all to maternal effects.

So why should organisms experience this? Is this a random process or might the maternal effects confer some survival advantage onto the progeny? We might expect that if such an effect had a deleterious action on the progeny, the progeny would be less able to survive with decreased reproductive fitness; then this would decrease the chances for the progeny to continue. Over time we might expect these individuals, and ultimately the population and species, to disappear. On the other hand, if such changes were to offer a survival advantage to the individual, we might expect them to survive these environmental disturbances initially perceived by the mother to a much better degree. At the time of birth, progeny will inherit their parent's environment. Environmental disturbances in the form of climatic changes, food source changes, geological disturbances, predation and overpopulation will typically occur over many generations, thus, if the progeny are prepared to anticipate these stressors they might be better prepared to meet the challenges of their inherited environment.

7.2   Epigenetic actions of stress on reproduction

Epigenetic effects have been demonstrated through studies of prenatal stress and maternal malnutrition. The care received by an infant early in life can also produce changes in the development of the neural system regulating the response to novel situations and social behaviour. Transmission of such effects can occur through subsequent generations because of changes in the alternation of reproductive and stress-associated behaviour to the offspring. As we have mentioned earlier, the more complex the organism, the more likely there are to be maternal and parental effects on the progeny.

7.2.1   Epigenetic factors of stress

In humans, for example, studies indicate that the majority of abusive parents were also abused as children, or those who grew up in less nurturing environments will tend to show less sensitivity to their own children. Studies also suggest similar effects among primates. A majority of rhesus monkeys, for example, display abusive behaviour to their own offspring if subjected to an abusive early development. This effect is significantly reduced if the abused offspring is fostered to a non-abusive parent.

While an abusive situation might be considered an extreme example, other studies both in primates and rodents indicate that reduced mother–infant contact can also lead to changes in the offspring's interaction with their own progeny. It is thought that such epigenetic effects on progeny have an adaptive effect. Although it is difficult to establish the significance of such adaptive effects in humans, a number of clues have arisen from studies on animals.

Most of the studies of epigenetic effects of maternal interaction have been performed in rodents, primarily in a laboratory setting. Because these species have a shorter generation time and produce a greater number of progeny they are particularly amenable to study in a laboratory environment. Moreover, although much less complex than primates and humans, the rearing and social conditions of rodents are relatively easy to manipulate in a laboratory context. A number of basic study situations have been employed.

All animals will display a set of behaviours expressing parental care unique to that species. For humans, we think of this as showing affection for the infant, expressed in the form of caresses, kisses hugs and play. Rats, on the other hand, demonstrate this through weaning, licking and grooming. The mother rat will be attentive to the pups and may retrieve a pup to bring it back to its litter if it has become separated from the rest. Female pups that are artificially separated from their mother for a sufficiently enough period show reduced or impaired maternal care toward their own offspring when they become adults as shown by a decreased tendency to retrieve separated pups or by reduced bouts of licking and grooming. This reduced care of their progeny then affects the maternal care displayed by the progeny when they too reach adulthood.

There are a number of physiological and behavioural effects that occur in both mother and pup during these bouts of licking and grooming. For example, licking can serve to stimulate the pups and modulate body and brain temperature. On the other hand, the mother can ingest and reclaim additional salt and water from the pups to replenish some of that lost through lactation.

Maternal deprivation of the young is a form of stress. The young are largely dependent upon the mother for survival and if there is significant reduction of these maternal behaviours it induces a stress response. Normally, the vocalizations of the pups will attract the attention of the mother and it will receive the proper care. If this does not occur, however, and the pup remains unattended and isolated, there will be a significant increase in the stress response, resulting in the activation of the hypothalamic–pituitary–adrenal (HPA) axis. In one set of studies, rat mothers (dams) were classified as high, medium or low with regard to their frequency of grooming and licking interactions with their pups. The offspring of low-frequency dams showed a much higher and prolonged increase in ACTH and corticosterone in their blood compared with pups reared by high-frequency dams. Such findings indicate increased secretion of corticotrophin-releasing factor (CRF), the neurohormone responsible for the activation of the HPA axis into the pituitary gland. Under normal circumstances, as described in previous chapters, elevated levels of the glucocorticoids, in this case, corticosterone, provides a negative feedback to the release of CRF from the hypothalamus. However, in pups reared by low-frequency dams there was a decreased ability of the corticosterone to inhibit the release of CRF. In other words, CRF was less sensitive to the feedback effects of glucocorticoids. Because CRF release was not reduced, the higher levels of CRF continued to stimulate the release of ACTH from the anterior pituitary gland, resulting in continued synthesis and release of corticosterone from the adrenal gland. As a result, the pups from low-frequency dams became more sensitive to the external effects of stress and maternal deprivation than those from high-frequency dams.

Numerous studies in laboratory rodents have confirmed the reproducibility of such findings, but similar results have been obtained in different species under very different conditions. In the studies previously described in rats, the females were selected on the basis of particular behaviours or artificially removed from the pups. In the wild, however, numerous environmental situations may act to reduce the time available for the mother to interact with her offspring and, in addition, she may be under greater stress. A species that has been studied in great detail is the snowshoe hare. In Canada's Yukon Territory, populations of this species typically undergo ten-year cycles in the numbers of individuals in the population. During a period of population decline, almost every individual dies due to predation. Thus, these animals are under considerable pressure to survive and reproduce. In a set of very interesting studies, Dr Rudy Boonstra and his colleagues at the University of Toronto, Scarborough, have shown that this predation pressure either alone or coupled with a decrease in food availability, increases the levels of ACTH, glucocorticoids and other indicators of the stress response in the affected hares. Moreover, they have shown both experimentally and under natural conditions that these higher stress levels can lead to smaller litter sizes and generally reduced reproductive fitness of the parents. Perhaps even more remarkable is that the progeny born to these 'stressed' parents show a higher level of stress response as indicated by activation of the HPA axis. The progeny show greater sensitivity to stress and exhibit less exploratory behaviour and increased wariness to potential predators. The researchers hypothesize that these epigenetic effects are an adaptation of the progeny during high environmentally stressful seasons which allows them to become more cautious and, therefore, more likely to escape potential predation.

7.2.2   DNA methylation of glucocorticoid and oestradiol receptor

Researchers then needed to ask how the effects of the parents could act to permanently change the behaviours of their progeny. The glucocorticoid receptor is one molecular mechanism that has received much attention. As you will recall from earlier chapters, the glucocorticoid receptor is the receptor responsible for, among other things, mediating the negative glucocorticoid feedback to the brain to inhibit the release of CRF. Recall also that the glucocorticoid receptor is a transcription factor. In other words, the activation of this receptor will directly affect the expression of various genes immediately after the glucocorticoid molecules bind to their receptor.

Glucocorticoids are members of the steroid family of hormones so are lipophilic, meaning that they are attracted to other fatty and oily molecules such as those found in cell membranes and can freely pass through membranes. With the aid of binding proteins in the blood, steroids can traverse the body to enter virtually all tissues in the body. Because of this, if a stressful situation is perceived by the mother the resulting glucocorticoids she secretes will pass into the embryo or fetus by diffusion across the blood–placenta barrier. The blood–placenta barrier is a cellular and molecule barrier that helps protect the offspring from noxious chemicals produced or ingested by the mother. It is not a perfect barrier and some molecules, such as glucocorticoids, perhaps as they were evolved to, pass freely through the barrier. Thus, depending on the intensity and duration of the mother's stressful episode, these molecules may pass into the progeny and affect their development. The maternal steroid hormones are identical to those hormones found in the progeny and, therefore, interact with the progeny's glucocorticoid receptors just as well as their own.

7.2.2.1 Glucocorticoid receptor.   Numerous behaviours associated with both stress and reproduction are mediated by the glucocorticoid receptor. We have previously discussed the role of this receptor with respect to the sensitivity to stress, but it is also associated with the cognitive abilities and response to reward. Licking and grooming behaviours exhibited by the dam are associated with variations of the glucocorticoid receptor messenger RNA (mRNA). The promoter of the glucocorticoid receptor gene contains sequences that are subject to methylation. Some studies indicate that elevated levels of licking and grooming in dams are associated with decreased methylation at these sites, leading to increased levels of glucocorticoid expression in the hippocampus. The result of this is that the dams are more sensitive to the negative feedback effects of circulating glucocorticoids and therefore the HPA axis is rapidly inhibited after exposure to a stressor. In other words, the glucocorticoid levels fall and do not remain elevated following the stressor.

Such studies predict that in situations of high glucocorticoid receptor gene methylation, less of the receptor is expressed and the brain becomes less sensitive to the negative feedback actions of the glucocorticoids and, as a result, circulating glucocorticoids remain elevated. However, because the glucocorticoid receptor is itself a transcription factor, this increased methylation of the receptor also affects other genes that the glucocorticoid receptor itself regulates. Thus, the effect just on this one receptor can have profound implications for a number of physiological systems and behaviours. And because this methylation can endure across generations, then offspring and grand offspring may also be affected.

7.2.2.2 Oestradiol receptor.   The glucocorticoid receptor is not the only gene that can be methylated. A number of studies support the presence of methylation on the oestradiol receptor. Oestradiol is a sex steroid that plays a role in sexual differentiation, reproductive physiology and behaviour of both sexes, although its actions are more profound in females. Rat mothers that lick and groom their pups more than most – known as high-frequency dams – have been found to have more oestradiol receptors in their medial preoptic area than mothers that do not lick and groom their pups so much (low frequency). The medial preoptic region of the brain plays a major role in the development of the neural circuits associated with differentiation and maintenance of the reproductive cycles. Methylation of the promoter in the oestradiol receptor gene is elevated in the offspring of low-frequency dams compared with high-frequency dams. Thus, this would predict that oestradiol receptor expression in offspring that have been subjected to certain degrees of maternal deprivation would be reduced and, therefore, less sensitive to the actions of oestradiol. This would further predict that in a population, widespread and chronic stress leading to a decrease in maternal care could lead to a decrease in reproductive fitness as the combined result of reduced sensitivity to oestradiol and corticosterone. These studies also reveal, however, that there are significant variations in the amount of methylation in these promoter sites among the pups in the litter. Thus the reader is cautioned that lower levels of maternal care do not necessarily lead to reduced sensitivity to oestradiol and glucocorticoids.

The majority of these studies have been performed in rats and the mechanisms governing the regulation of methylation and subsequent epigenetic-regulated physiology and behaviour in humans are likely more complex and are, on the whole, less well understood. However, a recent study by Michael Meaney, and his colleagues at McGill University have shown that increased glucocorticoid receptor regulatory region methylation is associated with childhood abuse in humans and is somewhat more common in suicide victims, which suggests a common mechanism of the epigenetic effects of methylation in humans.

7.2.3   Epigenetics of germline development

One theory suggests that the epigenetic regulation of DNA may have evolved as a prerequisite to multicellularity to allow the development of lineage-dependent gene expression in cells. For example, as each new cell lineage forms within the embryo, a unique pattern of silenced and expressed regions occurs. Another function of the epigenetic methylation of DNA may be to permanently silence 'junk' DNA. Junk DNA consists of repetitive sequences or DNA that has entered the genome throughout evolution primarily by viral transfection and, if left unchecked, could lead to instability within the genome and loss of reproductive fitness in subsequent generations. The demethylation of silenced genes appears to be required for the formation of germ cells, oogonia and spermatogonia during embryogenesis. This is not a perfect mechanism and errors in this demethylation can lead to defects in the germline. Increased incidence of these defects has been reported in some cases of assisted reproductive technologies, such as *in vitro* fertilization, but it is not known whether exposure to environmental stressors is associated with this. It is likely that in some species, exposures to environmental toxins may interfere with this process.

In mammals, widespread demethylation can also occur after fertilization just before and after implantation in the uterus, but again it is not known if high levels of maternal stress can affect this process. The process does vary considerably among mammals and in a number of non-mammals. In the zebrafish, for example, genome-wide demethylation does not appear to occur.

In all animals, the germ cell line is derived from somatic cells. As you may recall from Chapter 1, the germ cell line is unique in that it possesses a haploid set of genes, unlike the somatic cells which are diploid. In other words, the germ cells possess only half the number of genes and chromosomes that somatic cells have. This occurs in part because genes associated with the development of the germ cell line are silenced in somatic cells. Then as a result of the demethylation that occurs during the development of these germ cells, these genes are no longer silenced and a somatic lineage of cells begins to differentiate into germ cells.

In mammals, the primordial germ cells begin to form early in embryogenesis near the region where the kidneys form. As they begin to differentiate they start to migrate toward the coelomic cavity into a region called the genital ridge, which is the precursor of the gonads. Once in this region, depending on the genetic sex of the individual, they will differentiate into either the primordial egg cells (oogonia) or sperm cells (spermatogonia).

7.2.4 Epigenetics, imprinting and assisted reproductive technologies

Epigenetic regulation becomes particularly tangible when we consider that the environment influences the genes parents transfer to their children. Although the majority of offspring that develop via assisted reproductive technologies (ART) are healthy, the ART process bypasses a number of normal biological processes such as selective gamete reabsorption, selective sperm uptake and sperm competition and subjects the gametes to novel environmental stressors such as the presence of unnatural concentrations of hormones, culture media and physical stresses. Some studies suggest that low birth rates, congenital malformations and imprinting disorders in some ART offspring may be attributed to epigenetic variation.

Genomic imprinting is an epigenetic process in which the male and female germline of higher mammals (therians) induce a sex-specific action or imprint on particular chromosomal regions. Because of this, paternal and maternal genomes differ in selective regions but these imprints are required for normal development. The imprinted regions differ with respect to DNA methylation and histone modifications which affect the expression of certain genes. These genomic imprints are removed in the primordial germ cells that develop during embryogenesis then are newly established as a consequence of the new genome in the later stages of germ cell development. About 80 imprinted genes have been identified, many of which are associated with the regulation of resource acquisition in the embryo and fetus. Imprinting may have coevolved with the evolution of the placenta.

In the genetic conflict theory, the paternal genome has evolved to extract as many resources from the mother as possible. On the other hand, the maternally inherited genome protects the mother from being exhausted by the fetus because the embryo and fetus develop at the expense of the mother. Imprinted genes, for example, play a major role in the development and function of the placenta and regulate the acquisition of nutrients by the fetus by affecting the growth of the placenta through the regulation of certain molecular transporters and ion channels. A number of paternally expressed genes enhance fetal growth, whereas several maternally expressed genes may restrict it.

The best example of the relationship between fetal growth and parental interest is the IGF2 gene. IGF2 encodes insulin-like growth factor-2, and in the mouse it is found that if a lot of it is made the mouse grows bigger. The curious thing is that although the mouse gets one IGF2 gene from the mother and another from the father, only the one from the father (paternal) is active. If the IGF2 from the father is defective, the baby mouse is small, but if the copy of IGF2 from the mother is defective, it has no effect because it is not active anyway.

During some forms of ART, novel stressors may affect the pattern of imprinting and general epigenetic regulation. The nutrient concentrations found in the culture media used differ in many ways from what would be found *in vivo*. It is known that the availability of nutrients to the fetus and embryo differs significantly between mother–fetus pairs as a consequence of food available, the metabolism of nutrients and the transport of nutrients to the fetus. Despite this, most ART developed progeny are normal, attesting to the great developmental flexibility that is inherent in the progeny.

Some researchers have suggested that epigenetic alterations in DNA methylation in other genomic regions may occur as a result of different nutrient states, although is not known whether this can affect the germline or, indeed, whether it occurs in mammals. DNA methyltransferases and histone methyl-transfereases both use adenosylmethionine as a methyl donor. This substrate requires folic acid for its synthesis and the enzymes associated with it require vitamin B cofactors. Given that folic acid and vitamin B are obtained through the diet, then nutrients that are particularly low in these substances could have an affect on normal methylation. Currently, there is no strong evidence that genomic imprints can be affected by nutrition in humans. Nevertheless, it has been noted that babies conceived through ART are associated with lower birth weights than infants conceived without ART. The mechanisms for this are not clear but some studies have indicated that insulin-like growth factor binding protein-1 and some HOX genes may be implicated.

Some studies have investigated whether changes of the *in vitro* culture conditions of pre-implantation mouse embryos can affect the behaviour of adult mice. Although no changes in sensory and motor development were noted, effects on anxiety, locomotor activity and spatial memory were found. It is not clear, however, how relevant these studies are to human children born following *in vitro* fertilization. In fact, no studies have found any marked difference between children born following ART and those conceived spon-taneously. However, these studies have raised important issues in agriculture where ART is frequently used and may play a role in the reproductive fitness of future generations. For example, in livestock breeding programmes using *in vitro* fertilization the numbers of malformations, such as large calf syndrome, are much higher than those seen under normal birth processes.

7.3   Environmental effects on epigenetic regulation

Chemical and environmental toxins can alter DNA methylation patterns and result in altered epigenetic regulated physiologies and behaviour. During embryonic development the male germline is particularly sensitive to the environmental effects of chemicals and toxins. As mentioned previously, genomic demethylation occurs around the time when the primordial germ cells are forming. At this time, these pluripotent germ cells can differentiate into either oogonia or spermatogonia, depending upon the genetic sex of the individual. A complete demethylation of these cells occurs by the time these sex cells migrate to the primordial gonad. Then, during the period of sex determination in the gonad there is sex-specific remethylation of the germ cells. During this critical period of demethylation and remethylation the DNA is particularly sensitive to environmental toxins and chemicals which can result in altered DNA imprinting. These environmental effects could include endocrine disruptors (molecules that inhibit the normal endocrine circuits), heavy metals, abnormal nutrition, irradiation and chemotherapy.

7.4 Summary

Epigenetics is the science associated with the environmental regulation of gene expression. Epigenetic inheritance, therefore, consists of those traits that are inherited but cannot be explained fully by the genes alone. The basic concept is that environmental actions induce structural changes to the genes both in the germline and somatic cells of developing embryos. These changes affect the physiology and behaviour of the individuals as they become adults. Progeny can be exposed to early developmental stress directly, or indirectly by maternal stressors. Such stress can change reproductive physiology by modifying the expression of certain genes. In particular, DNA methylation of the oestradiol and glucocorticoid receptors has been implicated in increased sensitivity to stress and reduced reproductive fitness in progeny when they mature into adults. Some studies indicate that the epigenetic regulation of parental imprinting may be affected in some forms of assisted reproductive technologies.

8

Species in captivity: stress in agriculture and aquaculture and effects on habitat loss

The rapidity of change and the speed with which new situations are created follow the impetuous and heedless pace of man rather than the deliberate pace of nature.

Rachel Carson, *Silent Spring* (1962)

8.1 Introduction

As I write these words, my cat is purring, curled up on a blanket beside me – the apparent signs of a contented cat. She was acquired as a kitten after she was weaned and since then has not known any other life besides the one she has residing with us. She is healthy, well-fed and has the opportunity to go outside whenever she wants, hunt and generally do as she pleases. But is she living the life to gain the full experience of cat-hood? Does she care? Who knows? She cannot communicate such complex ideals to me. I assume she is happy because she does not show any behaviours associated with distress and she is beside me generally whenever I am around.

We assume that events that induce stress in us as humans are not that different from those perceived by non-humans. But this is frequently not the case and it can be challenging to determine whether an individual or population of any species is experiencing the early stages of a stress response. There are two elements that we consider when trying to determine the intensity of a stress response – objective and subjective. Objective responses are easier to quantify. Quantifiable responses are those that can be measured by various

Sex, Stress and Reproductive Success, First Edition. David A. Lovejoy and Dalia Barsyte.
© 2011 John Wiley & Sons, Ltd. Published 2011 by John Wiley & Sons, Ltd.

physiological and metabolic parameters, for example, glucocorticoid levels. As previously discussed, the similarity in the glucocorticoid response among numerous species gives us sufficient confidence that an increase in blood glucocorticoid levels is indicative of a stress response even if we measure it in a species that has not been previously studied. Similarly, if population sizes decrease, or if there are sufficient changes in body mass along with the presence of certain stressors, we can be reasonable confident that a stress response has occurred.

We might think of these measurements as objective in that a mathematical relationship between the stressor and the physiological response can be established. What becomes problematic is understanding the subjective experience of the stressor. As we discussed in previous chapters, individuals, not to mention species, have vastly different sensitivities to stress. We do not understand how to translate measured changes in metabolic and physiological parameters into the subjective experience of how an animal perceives this stress. Unfortunately, we generally cannot determine if an animal or species is under stress until there is a major change in their behaviour or, more frequently, population characteristics. Clearly it is not possible to sample the levels of stress indicators of all affected species. Because all species have evolved to adapt to various stressors during the adaptation to their environment, it is only when this adaptive ability has been breached that we can measure distinct attributes that indicate a stress response. Frequently then, by the time we can measure such a response, a population may be well into a period of decline.

Herein lies the anthropomorphism of the perception of the stress response. Anthropomorphism is the practice of placing human values on non-human species. Typically, the more we perceive an animal looking or behaving like us, the more likely we are to ascribe human characteristics to it. Common sense might dictate that anything we perceive as stress will be stressful to any other species. But as we have argued in previous chapters, species have adapted their stress response to their sensory experience of the world around them. For example, a species that does not have a visual sense cannot perceive the stress induced by light levels that are too high or too low. A large component of human stress is perceived or anticipated even if there are no events that will induce this stress. We imagine a myriad of stressors because of our developed mental processes. However, the human experience is distinct to humans. As humans, we can never understand the subjective experience of another species because of their unique sensory, processing and response ability.

Because of this, we need to rely on objective measures of stress to determine negative actions on a species. Sometimes stressful experiences are self-evident. If a species has evolved to a particular habitat and niche and it is then suddenly transferred to a completely different environment, for example, this is likely to induce a stress response. Of course this might be caused by the stress of capture and handling but it is also likely to be a result of elements of the environment such as temperature, range, nutrient availability and change, or population dynamics, for example.

A stress may be acute, resulting in a strong stress response upon the initial perception of the stress which subsides once the species adapts to its new environment. Such a stress is not likely to be injurious to the population and a proportion of individuals may adapt over time to the new environment. Because of the great variability in individual perceptions to stress, however, some individuals will experience high acute stress. In extreme situations this may lead to a decrease in reproductive ability or even death of these individuals. If enough individuals are affected like this within a population, then the continued loss of reproductively viable individuals ultimately may lead to a loss of the population fitness as a result of shrinkage of the gene pool. Sufficiently different environments may lead to more chronic stresses to which the animals cannot adapt and the population goes into decline. However, a decline does not exclude adaptation. In many cases, the population may recover as the result of a smaller group of individuals that do possess an adaptive ability.

8.2 Management of wild species

Such issues are important in understanding the human management of wild species. Humans have enormous impact on wild populations by decreasing their range, transplanting various species to new environments, selective breeding and domestication of wild species and selective modification of environments through the removal of specific species. Human society has probably had a greater effect on the survival on other species than any other one species in the history of the planet.

Biologically, every species lives at the expense of other species. If not, most species would not have evolved to a point where it could be established as a distinct and separate species. We recognize that it is important to protect as many species as possible, although we cannot protect all species on this planet. All species use other species for food protection and resources. Humans are not alone in this respect, and we recognize that in our current state of evolution, we will continue to use other animal species for food, protection, companionship, amusement, study and transport.

We have to face the reality of the increasing human population and our need for food and other biological resources (Figure 8.1). This impact has been particularly felt by the world's fish stocks. Global per capita fish consumption

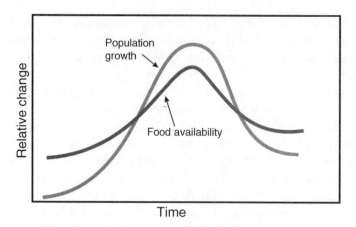

Figure 8.1   Relationship between population growth and food availability. Populations will go into decline once the need for nutrients is greater than nutrient availability

has doubled over the last 50 years, and it is projected that production would need to double again over the next 25 years to meet the needs of the increased human population. Unfortunately, most wild stocks are already heavily depleted, overfished and generally utilized to their fullest capacity. Recent studies have indicated that in a number of regions of the planet where the density of human population is high, the presence of large-bodied fish declines and fish communities become dominated by fewer and small-bodied species. Should this trend continue, and in the short term all indications suggest that it will do so, then we can expect the global aquatic ecosystems to evolve into directions different from what we have experienced in the past. The rapidity of human population growth and need for food and resources has outstripped the ability of these ecosystems to recover and remain in a productive equilibrium. Thus, management of the remaining stocks is essential if humans are going to continue to utilize these species.

   Management of wild stocks, regardless of the species, produces a number of effects both on the managed species and on other species of the ecosystem in which they are resident and, of course, on humans. There are several ways to approach these problems. Species may be managed in their natural ecosystem. However, although this might be preferable for the vitality of that species, for humans it becomes a costly affair with relatively low productivity. Moreover, the policing of such regions becomes a challenge. Alternatively, intensive agricultural and aquacultural programmes increase productivity for humans, and allow wild stocks the opportunity to recover. Again, this introduces a new set of problems through modifications of habitat and ecosystem that occur around regions of intensive farming as a result of structural changes, nutrient loads, fertilizers, disease and detritus production, to name a few examples. We might also consider diversifying to new food species to reduce the impact on the preferred species and animals in general. It is expected, however, that all approaches will be utilized to greater degrees in the future.

8.2.1 Stress and animal welfare

Human interaction with a species does not necessarily mean its demise. The domestic fowl, for example, by providing a food source for humans and by adapting to modified conditions has become the most common bird in the world. Nevertheless, as some authors have pointed out, the chicken, although successful in a biological sense, has derived far less benefit from humans than humans have from chickens.

This brings up the concept of stress and animal welfare. These terms are frequently used in conjunction, but have fundamentally different meanings. We have defined stress in this book as events that elicit a homeostatic deviation from normality. And, as we have pointed out earlier in this chapter, we have focused on physiological and resultant behavioural changes associated with stress. Animal welfare, on the other hand, might be defined as the animal's perceptual interpretation of the environment they find themselves in. This makes animal welfare much more difficult to measure, but with a biological understanding of the species that is affected we can probably make some reasonably accurate predictions as to what rearing environment would keep the animal's discomfort to a minimum. It would be naïve to think that all animals in captivity are held under conditions that reflect all of the animals needs. Sadly, I have had the opportunity to observe a number of agricultural, aquacultural and zoological operations where the resident species are not appropriately cared for and show signs of considerable distress.

That being said, we need to consider that many species used for economic purposes differ considerably from their wild brethren. Food animals, such as chickens, cattle and fish, are bred for characteristics that will provide the greatest benefit for the cost of rearing them. That is, the rearing conditions are optimized for reproductive output and nutrient protection with a certain amount of attention directed to the welfare of the animal. Are these animals under stress? In some cases they may very well be but in other operations they may not be. They have been selectively bred to survive under these conditions. In these conditions of captivity, there is considerable variability among operations. We cannot compare these individuals to a wild-type individual, as their gene pool and resulting epigenetic modifications differ. These animals are bred to the conditions they are to be reared under.

8.3   Species in captivity

Although many of the species reared under agricultural, aquacultural and zoological operations are bred for that purpose, that does not mean they are necessarily free of stress. It is perhaps not surprising that much of our understanding of the effects of stress and reproduction comes from studies on animals reared for agricultural purposes. Many of these species become model systems through which to understand the physiology of stress and reproduction that can be applied to other species (Figure 8.2). However, domestic animals in many cases have been separated from their wild-type progenitor species for countless generations. Through selective breeding these animals have become distinct from their original ancestors and we have to consider them as a separate group. In general, we select these animals on their ability to reproduce and to provide us with the greatest amount of nutrients for the cost that goes into raising them. For example, animals that grow larger and more quickly are selected over those lineages that reach an optimum size more slowly.

In Chapter 5, we discussed the problem of habitat erosion due to human activity and the subsequent loss of range and foraging grounds for many species of animals. A similar problem occurs with respect to agricultural operations. Out of economic necessity, most farms need to be located reasonably close to either a major population centre or a transport route in the form of

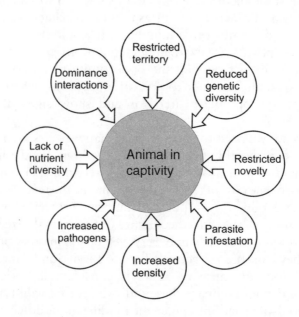

Figure 8.2   Some of the stressors experienced by animals in captivity

roads, rail or water in order that the animals and produce gain access to the market as quickly as possible. However, as human population increases, in many places, the urban sprawl encroaches on agricultural operations. In addition, the maintenance and operation of an animal farm is a laborious and costly affair. Thus, the size of these operations needs to reflect the economics of the financial return. Because of these constraints, farms are becoming smaller because there is less space available and it is more expensive to maintain larger, rather than smaller farms. Moreover, in the need to bring food to the population as efficiently as possible, combined with a need for the greatest economic return, intensive farming practices have evolved to keep the largest number of animals together in the smallest optimal space. An additional concern to animals is that they must be transported some distance to rendering operations. During transport, they may be subjected to a number of additional stressors, notably temperature and crowding.

To be sure, not all operations are of this type and depending on the type of farming operation, animals can experience different stressors. For example, some studies indicate that there is less HPA activation, as measured by glucocorticoid levels, in free-range chickens than in those held in high-density environments. Viral and bacterial diseases will spread much more efficiently when animals are held in high density. For example, sheep held in indoor pens are more prone to pulmonary adenomatosis, a progressive lung tumour that causes emaciation, weight loss and shortness of breath, than those kept in outdoor pens. Conversely, parasite infestation, for example, may be higher in free-range conditions. However, many diseases affecting farm stock can be treated with vaccinations and antibiotics, when available.

Aquacultural operations have problems similar to those encountered in terrestrial farming activity, with respect to overcrowding and disease. In both aquacultural and agricultural operations, excessive stress among the animals ultimately leads to poor meat quality, lack of the desired growth characteristics, egg production and reproductive potential.

Most species require a certain amount of space around them. The amount of this space is, to a certain extent, a characteristic of the species, and is a result of their evolution and adaptation to particular ecosystems. The sensory systems found within that species and the acuity of each sensory system impose limits on how animals interact with each other and, indeed, other species. Most individuals of complex species will seek out other members of their own species, but once the other individual becomes too close, they may seek a bit more distance. This is referred to as the attraction–repulsion relationship and it has considerable bearing on animals in captivity (Figure 8.3). For many species, territorial competition occurs. If the individual territory becomes too small, then aggressive behaviours develop as the individuals defend their spot. Alternatively, dominance hierarchies can occur, notably in birds and fishes,

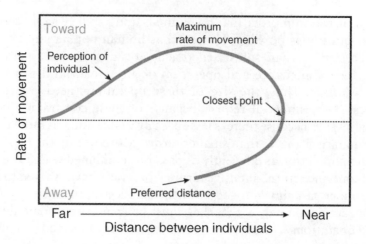

Figure 8.3 Relationship between attraction and repulsion among individuals

leading to the control of territories by dominant individuals. Subdominant individuals may experience numerous attacks by the dominant individuals, resulting in decreased health and death.

In both mammals and teleost fish, social interactions in small groups have been found to lead to elevated 5-hydroxytryptamine (5HT) levels and metabolism in subordinate individuals. 5HT is a critical neurotransmitter for regulating adaptive, cognitive and emotional processes. Atlantic salmon, when reared under high densities that typically occur during commercial aquaculture operations, show increased formation of dominance hierarchies. Although this effect may be reduced when food is abundant, many individuals continue to show increased 5HT activity and failed to grow even with the presence of excess food.

Keeping family pets is affected by fashion and commercialism. Animals are frequently bred for aesthetic characteristics rather than their ability to cope with the challenges associated with their life. Pet owners may perceive that their own understanding of the welfare of their pet is greater than the animal's perceptions. It is likely that the selective breeding of animals for any reason will affect some elements of the stress response. However, there are few studies to confirm this. It is usually impractical to perform stress-related studies on specialized breeds of dogs or cats. For example, although the public may accept stress studies on rodents and non-mammalian species, many people find it unpalatable to subject pet animals to stress studies. In addition, the breed itself may be so specialized that the results of the study may have limited application to other breeds or species. Finally, it may be difficult to procure sufficient animals of that particular breed that have been raised under standardized conditions required for controlled research studies. This is why we need to rely on studies of model animals. Most studies tend to focus on acute stressors, while animals in captivity generally have to contend with prolonged stressors. Ultimately we are dependent on the animal's behaviour and the incidence of overt manifestations such as neglect or disease to warn us that an animal is in distress.

Stress-related physiology is poorly understood in many of the species in zoological collections. Many of the species kept there are relatively rare, or difficult to capture and study, and again we rely to a large degree on individual observations. Any organism with a sufficiently complex brain, such as mammals and birds, will interact with its environment in similarly complex ways. Animals may experience exploratory drives, and be stimulated by novel situations and events in the environment. Although most responsible operations ensure that the cages, pens and ranges are as varied and close to the natural conditions as possible, in many situations they are not.

8.4 Summary

An animal's subjective experience of stress is difficult to ascertain and in many cases may differ considerably from a human's experience of stress. Moreover, many animals will perceive stressors that humans cannot perceive. For these reasons, it is often challenging to create the right conditions for animals in captivity. In order to ascertain the condition of animals in captivity, objective measurements are typically used to examine the health and behaviour of the animal. As the human population grows and there continues to be a loss of natural habitat for animals and agricultural regions for humans, farming operations will necessarily need to increase the intensity of their operations. This will lead to the selective breeding of new animal stocks able to adapt to these conditions. However, it is unlikely that new lines will be developed in the near future. Those that do will be bred and hybridized to produce animals with the desired economic traits.

9

A cellular understanding of stress and its relationship to reproduction

Did life lie groveling in darkness and sorrow, until the first dawn of the time-birth of things? Or what evil had it been for us never to have born? Whoever has been born must want to continue in life...

Lucretius, *De Rerum Natura* (first century BC)

9.1   Introduction

All cells have defence mechanisms of their own to protect them against stressors. With every organism, however, there are limits. Up until now, we have focused on the interaction between stress and reproduction in multicellular organisms. We will now consider the mechanisms found in individual cells. All single cells are organisms with their own stress defence and reproductive abilities. The cellular mechanisms in all multicellular organisms reflect the physiology of the original organisms that evolved before multicellularity evolved. However, multicellularity has limits. If cell division went unchecked, then we would end up with a sprawling morass of tissue which would eventually die. There would be no way to get nutrients to all parts of this organism. The evolution of more complex multicellular organisms occurred because cells differentiated into different functional groups that became tissues.

The size of the tissue is kept in check by chemical signals from other cells and tissues which act to limit its growth. A cancer, for example, begins as a cell system that has lost its constraints on growth and eventually its growth and need for nutrients overwhelms the rest of the organism. Thus, there are a series of molecular mechanisms in place to ensure that any one tissue does not

Sex, Stress and Reproductive Success, First Edition. David A. Lovejoy and Dalia Barsyte.
© 2011 John Wiley & Sons, Ltd. Published 2011 by John Wiley & Sons, Ltd.

dominate. In other words, reproduction of the individual cell is hindered by external events around them. In this respect, it is similar to any ecosystem. The overpopulation of any one key species could seriously affect the health of the entire ecosystem.

Reproduction in an individual cell, like that in complex multicellular organisms, reflects a trade-off between the need to reproduce and the need to survive. The basic mechanism is similar whether we are considering a single-celled organism or a cell that functions as part of a unit in a multicellular organism.

At the cellular level, stress defence mechanisms are associated primarily with maintaining molecular integrity. As we have previously alluded to, protection of the quality and integrity of our genes is essential to ensure that the individual is faithfully reproduced and the species continues. The DNA in our genes is constantly breaking, mutating and reconnecting. What stops our DNA degrading into a useless mess is the multitude of protective molecular mechanisms that have evolved to protect it from damage.

9.2 Evolution of cell stress, defence and reproduction

Before we consider these mechanisms in cells, it is useful to examine the theories of how the first cells evolved. In the early days of evolution single-celled organisms had to contend with the stressors in the abiotic environment and could not rely on the complex defence mechanisms found in multicellular organisms.

Let us consider the environmental stressors acting on a simple cell early in Earth's history well before the evolution of multicellular organisms. For life to flourish on this planet, these first cells had to contend with a unique set of stressors present in the environment at that time. Before the atmosphere possessed the thick protective coating it has now, it had a much thinner concentration of gases that allowed greater amounts of solar and cosmic radiation to infiltrate the atmosphere and bathe the planetary surface. Such radiation can break molecular bonds and induce new arrangements of molecules. It is so effective at destroying molecular interactions, that we routinely use radiation of various sources to kill pathogens. At some point in Earth's history, our atmosphere became oxygen-rich. It is not clear how this happened but when it did, it set the foundation for all modern organisms to develop.

The arrival of oxygen on the planet was a mixed blessing for the first organisms struggling to survive. On one hand, it can efficiently oxidize nutrients to produce an energy source for life. On the other hand, paradoxically, oxygen is toxic to all living things. In the presence of radiation or other energy sources, molecular oxygen (O_2) can be converted in a series of highly reactive oxygen species (ROS) such as peroxide (H_2O_2), superoxide radical (O_2^-) and the hydroxyl radical ($OH^{\bullet}$). These molecules can degrade the structure and function of numerous other molecular species, causing the disruption of cell membranes, destruction of proteins and degradation of DNA.

Today, aerobic organisms survive the high concentrations of atmospheric oxygen because of the oxygen-detoxifying systems present in the cell. Oxygen-intolerant species living today cannot survive in the atmosphere because they lack these detoxifying systems. As a result, these organisms are found where oxygen is absent or in insignificant concentrations. There are a number of chemicals produced by the body, or available in food sources, such as vitamin C, vitamin E and melatonin that help protect the body from the toxic effects of oxygen. Many of these chemicals may have had their origins around the time when the first organisms were evolving in the presence of atmospheric oxygen. As organisms became more complex, new proteins and enzymes evolved in response to the challenge of oxygen stress. These enzymes, including super-oxide dismutase, catalase and glutathione peroxidase, for example, appear to have evolved during this time as well. By mopping up the free radicals and converting them into harmless chemicals, the detoxification enzymes prevent damage to DNA or other proteins that are essential for cell function. Once multicellularity evolved, the regulation of these systems came under the control of the stress defence systems (Figure 9.1).

The formation of reactive chemical species continues to be a problem even today and they have been implicated in toxic effects of pollutants, cancers and the ageing process. A number of studies have indicated that ROS formation can cause damage to sperm and may result in loss of fertility. Higher levels of ROS in the testes correlate with decreased sperm motility. Under normal circumstances, seminal fluid contains a number of ROS-reducing enzymes, but as a function of inflammation, the ageing process or damage to the testicular tissues, higher levels of ROS may occur that overwhelm the detoxifying ability of these enzymes.

$$2O_2^{\bullet -} + 2H^+ \xrightarrow[\text{dismutase}]{\text{Superoxide}} H_2O_2 + O_2$$

$$H_2O_2 \xrightarrow[\text{Catalase}]{} H_2O + \tfrac{1}{2}O_2$$

Figure 9.1   Reactive oxygen species formation and detoxification. The superoxide radical is first converted into hydrogen peroxide and oxygen. Hydrogen peroxide is subsequently converted to water and oxygen

Another method by which the early organism could reduce the damaging effects of radiation was to time its reproduction to night time when solar forms of radiation are at their lowest point. This selection pressure to time reproduction to the dark hours may have been associated with the development of the first biological clocks. Biological clocks are ancient and are found in all kingdoms of life, including bacteria. The presence of clock mechanisms in all life forms attests to their physiological importance and these may have been among the first molecular pathways to evolve. In the simple organisms that populated the Earth in the primordial soup, evolving the ability to undergo DNA replication and cell division during the dark phase when ultraviolet radiation levels were at their lowest would have reduced the mutation rate within the DNA and hence increased the reproducibility of these organisms. A couple of billion years later, when organisms possessed the necessary cellular detoxification mechanism and the ability to move they did not necessarily need to have their growth phase in the dark hours. Instead, they could compensate in a behavioural manner by hiding in a protected space during the light hours.

Once biological clocks evolved, organisms became armed with a new defence system that allowed them to separate the day, and eventually seasons, into units. After the evolution of the nervous system the clock mechanisms became subsumed into circuits governing sensory-motor programmes and memory. The formation of these biological clocks led to the ability of organisms to predict disturbances in homeostasis.

9.3   Understanding cell death

Okay, let's fast-forward to the present time. I experienced temperatures lower than $-50\ °C$ on a number of occasions when I grew up. So when I moved to southern California at one point, I loved the warmth of the sun. As a young scientist with the need to publish the results of my research, combined with the desire to spend as much time in the sun as possible, I would frequently head to the beach early in the morning and work on my manuscripts. The sun was so warm, with the sand and water so welcoming. It was a wonderful feeling bathed in that warmth. A few days later, the warmth all but a memory, I was left with the itchy, red, peeling skin. Oops, forgot about the sunscreen and the resulting UV damage by sunlight. A number of us have had this experience even in the age of sun protection lotions and quite certainly our ancestors must evolved or learned to deal with the damaging radiation of sunlight as well. We know that too much sun is damaging to our skin and will lead to sunburn.

The peeling layers of skin are the cells that are dead as a result of UV damage. Those cells sacrificed themselves to protect deeper layers of the skin. So how did they die? First their DNA and protein was damaged by the UV rays. UV radiation can change how chemicals interact with each other. In the case of DNA and proteins, UV damage can break chemical bonds, leading to mutations in DNA and protein. Under normal circumstances, cells possess a number of proteins that remove and destroy abnormal proteins or those that have accumulated in too high of a concentration. But, if the damage is too great to repair, the cell undergoes so-called programmed cell death or apoptosis (Figure 9.2). We might consider this as being similar to cell suicide. This process is so controlled and orderly that once initiated, it cannot be stopped.

To best understand apoptosis we need to compare it with another type of cell death called necrosis. If the injurious factors impinging on the cell occur too quickly and too intensely, the cell cannot recover from such injuries. An example of the necrotic death would be a cut wound, at the edges of which the tissue turns an unpleasant grey/brown colour and then falls off as a scab. This type of death, in which the cell spills its contents because of a cut or leak in the

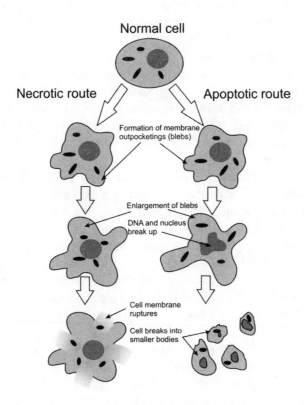

Figure 9.2 Changes in cell morphology as a result of necrosis and apoptosis

membrane, is necrosis. The cells are dead but the mess remains, including all the damaged proteins, acids and free radicals that are a risk to the surrounding cells. Somebody else has to come and clean that mess, otherwise it can damage and kill the surrounding healthy cells. In apoptosis, the cell self-digests and packages the leftovers into neat little packages that are not hazardous to the surrounding cells and can be picked up by the specialized cells of the immune system.

In sunburned skin, cell apoptosis is easy to observe, but what about other cells? Human cells do die, in fact thousands of them die every day under normal physiological conditions. Cell death can be caused not only by sunburn (UV) but also by gamma ray irradiation, toxins, heat or cold, lack of food and any unpleasant, stressful environment that does not support normal physiological processes or high or low salinity, for example. What happens when the cells find themselves in a stressful environment? Alarm bells ring, something is wrong, whether it is a damaged cell membrane or the DNA, the cell senses the damage and sets the evaluation in motion. Committees gather, meetings are attended, couriers fly from the damaged membrane, mitochondria or nucleus. The messages are conveyed through complex set of signalling cascades, something like a game of tag, where one protein passes the chemical tag to another protein after meeting in a special hub.

It is not all bad news. Although a number of molecular systems in the cell will be activated to induce a stress response, there are also a number of systems that are anti-death. So when a stress pathway is activated that may lead to the death of the cell, other systems attempt to block this mechanism. Although we might think of the pro-death mechanisms being the 'bad guys' and the anti-death mechanisms being the 'good guys', it does not always work that way. The pro-death bad guys are very important in development. In the development of a tissue, it is necessary that the tissue has limits to its cellular proliferation. For example, when human embryonic limbs develop they do not look like hands or feet, but more like fins, flattish outgrowths with web-like morphology between the digits. It is only later, when the cells that are in the positions between the digits are eliminated through programmed cell death or apoptosis, that the digits appear, eventually separating and looking like digits. You can think of this in terms of how buildings are constructed. First there may be a crane or scaffolding to erect the building, then when the building is completed, the supporting structures are removed.

Cell death is a very delicately balanced process. In the testes, too much apoptosis, because of stress or genetic abnormalities, causes infertility. Curiously, too little apoptosis could also lead to infertility because of the overcrowding of the seminiferous epithelium. It is estimated that between 50% and 75% of all potential sperm germ cells undergo apoptosis during various development stages; this is a normal occurrence of development. First, there are progenitor cells that lead the development to later cells. As these more mature cells develop, many of the progenitor cells will die. This occurs in all tissues, including tissues associated with reproduction. If this normal situation is interfered with, then the normal development of reproductive tissues cannot occur. Depending on the developmental state, related to maternal stress, epigenetic programming and the presence of toxins or lack of certain nutrients, then the cells' ability to react to these apoptotic signals during development can be compromised. This will lead to aberrations in the development of both the reproductive system and the various systems, such as stress, regulating reproduction.

9.4 Cell death and differentiation in reproductive development

The removal of excess cells occurs in the brain too. The sexually dimorphic nucleus (SDN), a group of neurons that lie deep in the mammalian brain area called the hypothalamus, is usually larger in adult males than females. However, male and female fetuses have the same sized SDN, containing roughly the same number of neurons. What happens to the neurons in the female? After birth, the male brain is usually washed over with sex steroids, testosterone, that gets converted into oestrogens, paradoxically that masculinize the brain, preparing the male brain for the needs of the adult male. Female brains do not receive the excess of sex hormones at that time. As a result, the neurons in the male SDN get the survival signal in the form of the sex steroid and keep living. The female neurons in the SDN lack this survival signal and some of the neurons undergo apoptosis.

In studies on rats, it has been shown that if the male newborn is deprived of the hormonal signal at this particular developmental window (just after birth), then the neurons of the male SDN die. On the other hand, if the female newborn is given the hormonal survival signal, a larger nucleus develops. But when the female is given either testosterone or oestradiol at this critical period, not only do the SDN cells survive, it also has the effect of masculinizing the brain as well. Normally, at this time in normal females, oestradiol levels are low and there are no significant amounts of testosterone. So this critical masculinizing period passes without consequence.

Hormonal manipulation of germ cells and other examples of reproduction-associated cell apoptosis are widespread. Just think how the female body changes during pregnancy. The growth of the uterus, placenta and breasts are all controlled by hormones. After pregnancy and lactation is completed, there is no longer the same hormonal input and the extra cells in the uterus and breast tissue are entirely dependent on the hormones. If these hormones are not present, then the cells have to die. This is probably a good thing too, as few women would like to look pregnant two to three years after giving birth.

9.5   Molecular mechanisms of cell death

So what happens when the cell is damaged beyond rescue? The pro-apoptotic factors battle with the apoptotic factors and the outcome is either life or death. Poisons such as UV, chemical toxins, ROS and X-rays activate apoptotic proteins, whereas survival signals in the form of growth factors and some hormones act on behalf of the anti-apoptotic camp. For example, progesterone can prevent cell death caused by ROS. The prime battlefields of these apoptotic and anti-apoptotic factors are the mitochondria. These are the gatekeepers of the apoptotic processes and are the chief decision makers in the death business. These tiny organelles that every cell has are the essential aerobic powerhouses of the cell. The theory is that way back when cells were evolving the mitochondria were originally a cell that could utilize oxygen for energy production. At some point an anaerobic organism ingested this aerobic organism to create a symbiotic organism capable of utilizing oxygen to metabolize energy-containing molecules.

When apoptosis is initiated, a signal is sent to the mitochondria to open the gates. Normally, the gates are shut and the protein machinery responsible for generating the energy currency, ATP, is kept inside. Once the gates are open, molecules such as cytochrome C leak out and activate self-digesting enzymes called caspases. Caspases are made in cells continuously, but remain in an inactive form until needed. In this respect, they are similar to the digestive enzymes that only get activated by stomach acid and not in the cells that they are made, otherwise they would digest the cells that produced them. Activated caspases attack the cellular proteins and DNA and chop them into bits and the cell dies.

The gates of the mitochondria, therefore, are particularly important. One group of mitochondrial proteins called Bcl2 shut the gates while Bax proteins open them. Let us consider what happens when there is no gatekeeper. If the expression of Bcl2 is increased or Bax is turned off, the cells become more resistant to stress-induced apoptosis. For example, gonadotrophins decrease the levels of Bax in the germ cells and protect them from apoptosis. Interestingly, in mice, if Bax is lost, the ovaries continue to function long past the normal fertility period. This occurs because, normally, many oogonia in the ovary undergo apoptosis and only a certain select few can further develop into oocytes. So with no Bax, mitochondria are less likely to open the gates and less apoptosis happens in the oocytes. More oocytes means a longer reproductive life is achieved.

9.6 Heat shock proteins in stress and reproduction

Some of the first molecules discovered to be fighting for the cell's survival in the stress response were the heat shock proteins (HSPs). As the name implies, these proteins are generated in response to heat stress. The name reflects the first studies that led to their discovery. However, many other stressors and toxins induce HSPs. The protective qualities of heat shock proteins come from their ability to nurture other cell proteins. Proteins in the cell or organism range from tough ones, like the keratin that forms hair and nails which is extremely resistant to physical and chemical damage, to delicate proteins that even the heat shock generated by mild fever can damage. A good example of what happens to proteins with heat exposure is the chicken egg – think of the difference between raw and boiled eggs and you can see the effect of the heat. Obviously heat shock proteins cannot restore chicks from boiled eggs, but they do have a very important function in maintaining healthy proteins under more normal conditions of heat stress.

Most mammalian testes are kept 2–8 °C lower than core body temperature by being outside of the body cavity. Higher testicular temperature leads to increased levels of heat shock proteins, which are designed to prevent and limit the damage to cells. Some damage still occurs with prolonged exposure to heat. Men who work in high temperature environments, for example bakers, welders or foundry workers, or those who may need to sit for prolonged periods (drivers) or wear restrictive clothing, have a higher risk of decreased fertility. It is not surprising, therefore, to find that if some of the heat shock proteins are defective or not present there is an effect on fertility. In mice, HSP70 and HSPa41 knockouts (meaning mice in which the gene has been taken out) have fertility defects and increased apoptosis in testes.

9.7 Relationship between cell division and stress pathways

Different types of cells are subject to different amounts of stress. In general, dividing cells are more sensitive to apoptosis than cells that are not actively dividing. Because of this, even small amounts of stress to dividing cells can cause their demise. In contrast, cells that make up bones or muscle, for example, are post-mitotic cells, and can survive much higher levels of stress. Clearly, if your muscle cells were to start dying from just flu-induced fever, your chances of long-term survival would not be good. So the main difference between cells dying easily and those that are more resilient depends, in part, on whether their DNA is being replicated.

Some of the cell types in our bodies are particularly metabolic active. Brain cells, for example, require vast amounts of oxygen for normal function. Sperm cells have particularly high levels of proliferation and, therefore, also require a huge amount of energy. Obviously, these two cell types differ in a number of ways. In humans, the great majority of brain cells do not divide after the brain reaches its peak in development around late childhood. In general, this is true for all species of vertebrates, although less complex vertebrates have a number of regions of the brain that are capable of producing new neurons which can become associated with existing neural networks.

Most of the cells in the human brain are there for life and most of the cells that die will not be replaced. Interestingly, it remains controversial as to how much of a regenerative capacity humans have, because it is not practical to study cell division in living human brains. Sperm cells, on the other hand, are almost the complete opposite in terms of their proliferative ability. Human males can produce a few hundred million sperm cells per day every day for most of their adult lives. This requires a large amount of energy provided by the intake of nutrients and oxygen.

In the nucleus, the DNA – keeper of the genetic information – needs all the protection it can get. Germ cells especially need to protect their DNA, even more than somatic cells, as it might get passed onto the progeny, and any errors could lead to progeny with decreased survival abilities. Usually, if the DNA gets damaged in the cell special dedicated proteins can repair it. When the DNA is in the course of cell division, the repair is much more difficult to accomplish. When the house is on fire, the unpaid bills do not seem that important. And why risk any mutations? With 200 million sperm cells produced each day, it makes sense for human males to get rid of the ones with the damaged DNA or proteins and let the healthy ones go on.

In contrast, females start with only about 7 million oogonia. Because all DNA synthesis occurs in the embryo, this makes the sensitive window very short.

So how is decision reached – to die or not to die? That depends on the damage. Touching a boiling hot pot with your bare hands will certainly result in some cell death, but it is unlikely to kill you (the entire organism). Other cases are not so clear cut. For example, a moderate dose of radiation to a male animal will cause a reduction in sperm count but only after two months, as the most sensitive cells are the spermatogonia. As long as there are post-spermatogonial cells maturing, the sperm count will not be affected. Higher doses of radiation would cause sperm production to cease immediately, shutting down the entire reproductive system.

Apoptosis is used not only to get rid of unwanted cells in the development and faulty germ cells, but it also plays an important role in keeping cancer in check. Cancer cells, by definition, are faulty, and eliminating them by cancer drug treatments works in part because they are sensitive to apoptosis. However, when cancer cells become more resistant to apoptotic processes, their growth rate increases and, therefore, more aggressive medical intervention is required to treat the cancer.

9.8 Summary

A number of harmful factors induce a cellular stress response which can lead to cell death or apoptosis. This kind of apoptosis, as a pathological response to toxins, is important to eliminate defective germ cells. Physiological, or normal, apoptosis plays a role in organismal development and the regulation of hormone-mediated growth of reproduction-associated tissue, for example the uterus in pregnancy. Moreover, cells have also evolved a set of molecular mechanisms to protect them against injurious stressors to enable them to survive. Pro-apoptotic factors are important during the differentiation of tissues during development and play a role in the sexual differentiation of the brain and other tissues. Anti-apoptotic factors are also important during development, but also help protect the germline from destructive conditions. Under normal physiological conditions cells face a number of different stressors. The heat shock proteins and reactive oxygen species (ROS)-detoxifying enzymes are two such systems found in all cells that protect against a number of stressors.

10

Stress and reproduction in human society: implications for the twenty-first century

> The fall and recovery of populations, the politico-economic unification of human affairs, may present phases of intense stress and tragedy, periods of lassitude and apparent retrogression, distressful enough for the generations that may endure them, but not sufficient to prevent the ultimate disappearance of misleading traditions and the dominance of a collective control of human destinies.
>
> H.G. Wells, Julian Huxley and G.P. Wells, *The Science of Life* (1931)

10.1 Introduction

There are places on the Canadian prairies where you can stand and see fields from one horizon to another in all directions. The surrounding land is completely flat, with the vista broken only by a distant tree. This sense of openness is important to me, as I have discovered, after trying to live in a number of large cities. Despite having visited some of these large cities, such as New York, Shanghai, Mexico City or Sao Paulo, for example, I cannot imagine what it would be like to live there. In fact, because it was so difficult for me to live in the congested downtown of a large inner city, I changed my career to move to a new city. Now I live in a small town but commute regularly to work in Toronto. It is a large city, but manageable. Ironically, having numerous friends in these cities, their frequent comment is that they could not live on the prairies because they would find all that space so intimidating. The point here is that we adapt to our surroundings as we grow up.

Sex, Stress and Reproductive Success, First Edition. David A. Lovejoy and Dalia Barsyte.
© 2011 John Wiley & Sons, Ltd. Published 2011 by John Wiley & Sons, Ltd.

For a few millennia the world was a pretty spacious place, but in just a few hundred years the living environment for humans has changed drastically as our population has increased. Today, more and more of us live in highly urbanized centres. In this, our situation is perhaps not that different from many of the animal species we discussed in previous chapters. Humans have to contend with the effects of density, social interaction and nutritional changes just as much as other animals. These environmental changes are occurring at a much faster rate than the pace of evolution. Despite our enormous adaptive ability, there are signs that these changes are challenging our ability to cope.

10.2   The unique biology of humans

Humans are special – just as every species is distinct from other species. And, like other species, we possess a set of attributes that make us distinctly human. Part of our uniqueness stems from the complexity of our brains, and the resulting intelligence, but also the role that psychology plays with decision-making and survival. Although in some situations we can separate psychological factors from physiological factors, ultimately, our cognitive processing is related to our interaction between internal and external events. Our perception, therefore, plays a great role in affecting reproductive potential. Our perception of the world around us has to keep pace with changes in that world.

Humans are also unique in that we possess an ethical and moral obligation to divert our own resources to the planet with the goal of maintaining not only our own civilization but also other species and ecosystems. Most of us pay taxes to support our various levels of governments. We possess a direct benefit in the form of better education, recreation and health services, security, and better living conditions in the form of energy, sewage and water treatment for example. But many of these benefits are intangible to us. Our taxes may also be directed towards various regions around the world in the form of international aid, long-term ecological and exploratory studies and cultural exchange programmes. There is no direct benefit of such programmes to most of us, but ultimately part of the work we do benefits the planet as a whole. We do this as a society with the goal of maintaining future resources for us and our progeny – it's a sort of social and ecological retirement savings plan.

However, infant mortality rates are generally decreasing worldwide and people are surviving longer. We can expect this trend to increase for the near future, leading to a population density that will increasingly challenge the resources that we need to continue our lifestyle. Although overpopulation in a confined environment (i.e. the planet Earth) is not uniquely human, our ability to socially decrease mortality and extend our lifespan is. Life is sacred to us and currently we will do whatever we can to maintain our life. Humans, like all animals, have evolved to have a finite lifespan. A lifespan of a species reflects the time required for maturation and the production of offspring that ensures that the species continues. It also reflects the need for individuals to die off and be replaced by progeny to allow biological adaptation to the environment leading to evolution of the species.

It is not clear what the first evolutionary adaptations were that early humans obtained that led us down a path of developing independence from the original environmental conditions we began in. However, it is likely that in our current trajectory of development we will become less dependent on those conditions that we adapted to during the course of our early evolution, and more dependent on the conditions we have created for ourselves. Thus humans, more than other species, need to contend with stressors from our own making. If we do have longer lives and prolong our reproductive potential through technological means, the number of individuals living at the same time will increase. This will serve to increase our population density and put greater pressure on our limited resources. This will foster increased stress and hinder our reproductive potential.

10.2.1 Gestation and development

Human society is becoming more complex for a number of reasons. Greater complexity means that it takes longer to prepare our young for life in our society. When I was a child, school began at six years old and was completed by about eighteen years old. Most of my friends in high school found work immediately afterward. My son will begin formal education at four years old and will probably require significant university or college education to prepare for work and survival in our society. However, even before he begins formal education, he has already experienced a long gestation and period of parental interaction.

A relatively long childhood means that there is a longer window in which stress can affect the development of the reproductive system and the collection of behaviours associated with it. The mother–infant interaction period is lengthened and so potentially increases the effect of maternal stress on the infant. Humans are not entirely alone in this respect. Apes and Old World monkeys also exhibit long-term relationships between mothers and their juveniles. There are many advantages to such a relationship. A longer relationship with parents protects infants and juveniles from the risks of predation, accidents and social aggression. The extension of this protected early life stage in humans allows additional time for the interaction of the body with the physical environments. These interactions ultimately act to modulate the developmental process. Thus, for humans the personality is a product of experience. This is a feature that is far more exaggerated in humans than in other species.

There is, nevertheless, a compromise in terms of reproductive potential. The trade-off for human reproduction is that a longer juvenile period with increased parental investment leads to fewer progeny. In species with little or no social interaction, much greater numbers of progeny are produced with much reduced parental investment. There are exceptions of course. A number of live-bearing sharks have extended gestation periods, sometimes as long as almost two years, as well as an extended maturation period. But this is associated with a relatively low metabolic rate.

In humans, the rate and extent of neurological development is a particular feature. Brain growth is accelerated so that maximal brain size is reached well before maximal body size. During accelerated developmental periods, the circuitry of the brain is particularly sensitive to stress. As discussed in previous chapters, because humans have particularly complex brains, with an extended gestation and maturation period, the opportunities for stress-induced changes are increased. Having said that, humans also have an increased ability to deal with environmental stressors, and this, in part, plays a role in our adaptive abilities.

Gender development is an important part of early childhood development. It is a complex process in humans and, given its importance in our society, surprisingly poorly understood. Despite the complexity of the genetics of sexual differentiation, much of our gender development is social. Today, our definition of 'man' or 'woman' is frequently a social or legal definition that may not reflect our genetic inheritance. However, such situations are relatively rare, and in general society places vast importance on which gender we are. Our interaction with others is dependent upon our perceived gender and some of the social stressors that impinge upon us are, in part, dictated to us by the gender that we present. Regardless of our genetic sex, we may adopt the behaviour and mores of the opposite sex. We do this because of the genetic and behavioural plasticity inherent in our makeup as humans. However, social rules and perceptions are particularly important to define gender and constrain gender stereotypes and, sadly, cross-gender behaviour, while tolerated, remains poorly accepted in most societies.

For many species of primates, including humans, the differences between the sexes at birth are not obvious, with the exception of the external genitalia, and manifest only at later stages in life. Although like many other species, there is a tendency for males to be larger, the growth trajectories are similar until adolescence. This means that the larger sex (usually males) has to grow at a faster rate for a period of time and/or takes a longer time to reach maximal size. Moreover, in humans, females tend to reach sexual maturity at a younger age than males. This ultimately means that the window of growth and development differs between males and females. Because of this, environmental stressors that can affect growth differ between the sexes. Parental investment in humans is particularly high in order to protect and prepare their children for complex adult life. Moreover, because of the differing developmental physiologies between boys and girls, the amount and type of parental interaction will differ. This will ultimately affect how stress is perceived by children of different sexes.

Physiologically, human males and females are quite different and, as a result, the nutritional requirements differ. For example, humans are unusual among primates and mammals in that they have proportionately more body fat. In general, women have about 25% body fat relative to their body mass, and men about 14%. Humans differ even from primates in this respect. Female macaques, for example, have 13% of their body mass devoted to fat, whereas males have about 9%. The reasons for this are not entirely understood. Classically, it has been thought that fat reserves provided a nutrition cushion to protect the mother and infant during periods when there was a need for greater energy reserves, perhaps during long-distance travel for foraging, when only low nutritional food was available, or times when food was scarce during periods of migration or escape from predators. Some theories suggest that the conspicuous fat deposits in the buttocks and breasts in girls evolved as a signal for sexual maturity.

The rate of growth of adipose tissue during male adolescence lags behind that of the skeleton and musculature so that boys typically become leaner in their teens and accumulate adipose tissue relatively slowly during manhood. In macaques, 40–45% of the body weight is devoted to muscle. In women 36% of body mass is muscle and men have about 43% muscle.

Optimal amounts of fat in women have been implicated with normal pubertal development and the maintenance of menstrual cycles. Disruption of the normal development of adipose tissue can lead to a number of reproductive problems. The combined effects of low body fat, aberrant nutrition and excessive exercise have been cited as reasons for menstrual dysfunction in some female dancers and athletes. Such issues have become more important as society imposes certain ideals upon us that stray from the biological norms to which we are adapted through evolution.

In primates, the infant remains with the mother while she continues to interact with others of her species. This allows the infant to acquire an 'instant 'social network through its mother. This social network consists of both sexes and all age categories. Thus, as the infant matures, it expands both its social contacts and associated behavioural network. Disruption of the mother–infant bond, and hence this mobility, may disrupt the development of the normal behaviour of the infant as it matures into juvenile or adult stages. Under normal conditions this relationship to the mother in society allows it to establish its position in the social order.

The development of a society improves reproductive fitness in a species in a number of ways. Social living offers protection to lactating females, their infants and vulnerable juveniles. However, social interaction requires time and energy to maintain. Thus, the time the mother has for feeding or resting may be increased by her infant spending time with a group member 'baby-sitter'. In human terms, in some modern Western societies there may be little family support, for example, when both parents are working and the mother has less time to spend with the infant. Frequently the early socialization of the child is taken over, in part, by childcare facilities which may act to modify the socialization process by that parent and place it in the care of sanctioned societal norms.

10.2.2   Sex reproduction and fertility

It has been suggested that our own sexiness and need to thwart our competitors for a mate is the primary drive for the development of our intelligence. Such theories are based on the evolution of traits as being essential to increase our own reproductive fitness. If so, we would expect that some of our own stress is ultimately related our need to maximize sexual attractiveness. In human society, intelligence is one of the traits that increases our chances to reproduce. Perhaps like elaborate tail feathers in birds and antlers in ungulates, intelligence is a way of increasing our reproductive potential. Intelligence has developed in synergy with the complex living conditions of our society. Greater intelligence has likely led to an increasingly complex society, but this society demands and selects individuals with greater intelligence. But as we have argued in this book, our continuation as a species is dependent not only on our ability to procreate, but also our ability to survive in a hostile environment.

Society imposes a number of desirable characteristics upon us. We strive to be intelligent, sexually attractive and maintain a long healthy life. Few of us relish the prospect of age and losing our youthful vitality. As individuals, we are reluctant to accept the idea that this ageing process and eventual death is required for species and, indeed, social evolution, and as a society we have made great progress in the availability of high-quality nutrients and our ability to treat and cure numerous diseases and infections.

This increased level of nutrition and health has likely contributed to changes in our own maturation process. For example, there has been a decline in the age of puberty in human females, particularly in Western societies. Although the reasons for this are not entirely clear, it is thought that better nutrition, lack of disease and greater safety are implicated. Whether this effect will continue as humans face new challenges is not clear.

Although this argument makes a certain amount of sense, it does not necessarily hold for males. There is evidence that the sperm count of Western men has fallen by about 50% between 1940 and present. This finding is controversial and it is not clear what the mechanisms are. Changes in the definition of what meant by low fertility, or a lack of understanding of what constitutes normal levels of sperm have been cited as confounding factors in determining whether falling sperm counts in human males are a real phenomenon. One of the problems in ascertaining the reasons for this is that ejaculated sperm is several steps away from the site of proliferation and development of sperm. Although it is tempting to speculate that environmental and social stressors may be associated with this decrease, there is no clear evidence that this is the case, at present.

Another aspect to consider is that the traits that make Western males attractive to their partners today are not necessarily the same as those that were required, say, fifty years ago. In many cases, for example, potential male partners may be deemed attractive because they have child-rearing skills or are more family oriented, and not necessarily because of their sexual robustness. Thus, although some studies do indicate that sperm levels may be falling, and that the population of a number of Western societies has fallen, this does not imply a cause and effect. In other words, there is little evidence to indicate that populations are falling because of men's low sperm counts.

However, the amounts of stress impinging on children and adults are significantly different. We try to protect our children from as many stressful events as possible, while at the same time taking on those additional stressors. As parents, we spend a large amount of our time teaching our children to live in the society to which we are born. The acquisition of social abilities is often of greater importance to future reproductive success than the physical state or age at sexual maturity as predicted under conventional life history theory. Moreover, since the early 1960s, contraception has become increasingly widespread and accepted by society. Because of this, in societies where there is greater access to education and parental input, birthrates are either stable or in decline.

10.2.3   Social interactions

The per capita consumption of world resources is far greater in developed societies than in other areas of the world. Moreover, the rate of indigenous population growth in most developed societies is either stable or even in decline (i.e. not taking immigration into account). Some of this decline results from social trends, such as smaller family sizes, delaying pregnancy until later in life, thus delaying the time available for additional children, or choosing to adopt. In addition, because of the complexity of our societies, our work and career requirements have become increasingly complex, leading to less time for the rearing of children. Indeed, many couples choose not to have children. And herein lies an interesting question: do our career choices reflect a stressor upon us in that hinders our desire to have children?

Another aspect to consider is the number of women engaged in demanding careers. During pregnancy and birth, the amount of time a woman needs to spend away from her work and career is much greater than that for a man. Over the last century, far more women have taken on careers in which progress can be severely hindered should she want to take time off for pregnancy and child-rearing. The trend for increased numbers of women entering the work force may reflect the economic need for both individuals in a couple to work. Or on the other hand, it may also reflect women's enjoyment of having an intellectually and socially stimulating career, in the same manner that men have enjoyed for many generations.

Regardless of the reasons for lower birth rates, changing population dynamics can affect the stressors in a society. Increased lifespans coupled with lower birth rates induce an overall increase in the age of a population. An ageing population puts greater strain on the child-bearing group of society by diverting more social resources to the elderly. In many ways this is a good thing. By devoting more social energy to the elderly, we free some of the individual energy required for caring for the elderly. A healthy elderly population benefits society in terms of education, cultural and family continuity. However, we might liken society to an organism in that if more resources are required to deal with one situation, the resources available for other situations are reduced.

One of the major concerns that has arisen in a number of Western cultures is how to deal with the huge number of 'baby-boomers' that are now approaching retirement. The baby-boomers was a term given to the children born between the end of the Second World War and the late 1950s. This group of individuals are now in or approaching retirement and there will be an unprecedented amount of resources in the form of pension and healthcare that will be required to support them. At the time of writing this book, it is not clear that many Western societies will have the resources to adequately support this group of people.

In human society, money translates to energy. The more energy we have, the easier it is to overcome problems, right? I wish it was that simple. A very few of us may win millions in a lottery and can devote the rest of their lives pursuing the things they enjoy the most. But for the vast majority of us that is not the case. Couples are expected to work harder and longer. A greater financial income is usually associated with greater work complexity and greater stresses upon us. In some studies of job satisfaction, it has been shown that it is not the amount of wages that are a concern to workers, but rather a more enjoyable workplace environment. In Western societies the desire for greater wealth leads to people working harder and longer. This desire for wealth comes from many sources, including peer-pressure, media attention and aggressive marketing campaigns.

10.3 Stressors in human society

We have already discussed the prediction of increased stress on human society as a result of population increase and the increased competition for resources. In this respect human society is like most populations of any given species. But what actually happens during that population increase that acts to have a negative stress on the individuals living within it? A number of stress-induced conditions are on the rise in human society. An estimated 340 million people worldwide and 40–60 million people in the United States suffer from depression and anxiety disorders, according to the World Health Organization (WHO). Patients suffering from depression often also experience anxiety disorders. This co-morbidity occurs in 60% of cases. Stress has long been associated with the onset of mental disorders. Prolonged stress and history of depression is also implicated in Alzheimer's disease and other neurodegenerative conditions in the elderly.

Anxiety disorders affect 10–16% of the general population. The most common is generalized anxiety disorder, followed by social anxiety, panic disorder, obsessive compulsive and post-traumatic stress disorders. These conditions severely effect normal social interaction, work productivity and the ability to care for others. The development of these neurological disorders is complex, involving a number of genetic, social and environmental conditions. The incidence of depression, in particular, appears to be increasing in human society and particularly within highly industrialized societies. It has been attributed to increased social interaction, technology and information processing, for example. Paradoxically, it has also been associated with social isolation as more individuals cannot cope with the increasingly complex demands of modern society, and therefore limit their interaction with others.

10.3.1 Work, wealth and social interactions

A perception of safety and wealth leads to a reduction in stress levels whereas a perception of danger and lack of safety naturally leads to an increase in stress. Although our society protects us from a number of stressors, ironically, many stressors occur as a result of society. Social stressors are those that are attributed to interactions between individuals living within a society. Thus, as the complexity of a society increases so do the types and nature of stress.

Urbanization is becoming increasingly widespread world-wide. Not only are the number of individuals on the planet increasing, but also population centres are becoming larger and denser. These high population density areas put greater strain on resources, leading to greater competition between individuals and increased demand for jobs, training opportunities or finding a home in a desired neighbourhood, for example.

However, greater competition for such resources also means that more desirable aspects of society, such as trasnportation, our homes or vacation time, for example, become more expensive. Just as nutrients are energy for the individual, money serves as the source of energy for human society. If money translates to energy in our society, then the greater amount of energy is required to obtain the finance in order to afford such things. Thus, in an increasingly competitive society, work takes on a greater role in society and often both parents need to become involved to maintain what is considered an acceptable standard of lifestyle.

10.3.2   Perception of time and time constraints

Time is important. Most animals have a sense of time that relates to season or photoperiod, or the timing of tides, for example. Time allocation studies on female animals have helped to estimate energy budgets during different reproduce phases. Female mammals will adapt daily or seasonal activities by increasing foraging time and food intake and/or consuming higher quality food by entering feeding areas first, taking priority in feeding or having access to particular feeding places. Humans, however, have a much more precise awareness of time through the invention of mechanical clocks.

Since the invention of clocks and the subsequent division of time into smaller units, there has been an increasing tendency to complete tasks within a given time-frame and to assign priorities to tasks based on a sense of time. Time became a way to organize our day and coordinate our activities with other members of society. But time also becomes a limited resource. In modern societies work-related functions take on a time component. Because we are beginning to enter a period in our evolution where job tasks become more complex and resources more scarce, we are spending less time devoted to child-rearing. A number of studies also indicate that we are spending less time asleep. Extreme sleep debt has been associated with a number of mood and reproductive disorders.

This then increases the probability of parental stress in general as a result of the loss of attention and parental-associated learning and affection.

The concept of time has perhaps given us unwarranted confidence in our own ability to plan, and not consider the unpredictability of the future. Thus, reproductive events, such as the decision to have children may be relegated to a future time only to find that it is an inappropriate time or that time is not available, or even that our desire to have children has waned.

10.3.3 Information

The Internet has profoundly changed our view of the world. In a few seconds, after a few taps on the keyboard from the comfort of my own home, I can find out what is going on in virtually all parts of the world. When I was a graduate student and needed look up published reports, it would take me days of searching in a library for reference material, then all too frequently I would have to order the book or reprint and wait several weeks to read it. Now, I can do in about an hour what used to take me a few days. I can also access hundreds of radio and television stations from around the world and download podcasts from the Internet. I have access to far more information than I could possibly use. However, I am aware that there is much more information out there that I might need, but I do not have time to access it, much less read it. Information overload can be a stress. Humans have a great capacity to process complex amounts of information, but at the rate at which new information becomes available many people are finding it difficult to adapt.

At the time of writing this book, the world continues to undergo a severe market and financial crisis. Many governments have opted to divert cash resources to select companies to keep them solvent. This is one of the advantages of a society. Moreover, an efficient media system, in the form of newspapers, radio, Internet and television, keeps the population informed. Although this is advantageous to keep everyone informed, it has also led to a fear that workers will lose their jobs and an awareness that the security they thought they had for the future may not be there. Consequently, it has already had the effect of changing spending habits and retirement plans, and will probably affect many young couple's decisions to raise families. But as we have pointed out in previous chapters, reproductive potential and the perception or incidence of stress will be optimized to reflect both the ability to raise families and the ability to handle the limitations in the environment.

10.3.4 Industry and technology

All animals ingest nutrients, extract as much energy and raw materials as possible for growth and metabolic needs and excrete the waste products that cannot be used. Humans are no exception. In fact with our technological needs combined with our increasingly large population's demand for raw materials, we are churning out waste products on an unprecedented scale. This production of waste affects virtually every physiological system in our bodies. Pollutants are the waste products of a species and are a consequence for any species whose population has become unchecked. Pollutants occur in the form of excess heavy metals, sewage, radiation, garbage, chemical waste and noxious gases. Excess sound and light may also be considered pollutants. We are much better at producing such waste products than we are at removing them and, consequently, they build up over time. Exposure to these products ultimately affects our metabolic and cognitive abilities, as well as interacting directly with our reproductive physiology.

10.3.5   Disease and ageing

The combination of human over-population with declining living space for wild populations brings humans and animals into closer proximity. This has a number of effects on human populations. Human diseases such as viruses will be transmitted much more efficiently between human populations – especially in light of the ease of travel between remote parts of the world with major population centres. As these conditions also promote stressful conditions, this stress lowers the threshold for infection, thus exacerbating the situation. The closer proximity of animals and the loss of agricultural land also increase the probability of disease transmission from animals to humans. These conditions further reduce reproductive capacity through disease and pathology stress.

In humans, mental illnesses associated with stress and anxiety are increasing at an alarming rate in Western industrialized nations. Women are twice as sensitive to stress and anxiety than men. Psychosocial stress due to family, work and health concerns and physical stress can induce a variety of anxiety-related illnesses. Moreover, women tend to have longer episodes of stress and anxiety, and have a lower rates of spontaneous remission. The World Health Organization reported in 1990 that major depression is the single largest cause of morbidity for women and the leading cause of disability world-wide. Depression and other anxiolytic disorders affect women predominantly in their child-bearing and child-rearing years and may be caused by a combination of cultural and economic factors. Currently, women are still more likely than men to be dependent on their spouses for insurance and financial resources, to work part time, or to provide work for small businesses that do not provide health insurance. Thus, women are more susceptible to disruption of these resources through divorce, death and job loss, although there are some indications that these situations may be changing.

At present, in Canada and United States there is considerable attention on the increased incidence of obesity among the population. Numerous reasons have been cited for this problem, although many of the reasons ultimately are related to social stressors. Much of the type of work we engage in in urban society does not require much physical labour. Among white-collar workers most daily activities involve walking short distances, then spending excessive amounts of time on computers dealing with reports, emails, or spreadsheets, for example. Even among blue-collar workers, much of the physical labour required in the not-to-distant past has been supplemented by the development of labour-saving devices and machinery. Television, DVDs and computer games are now the recreation of choice for most individuals.

Our eating habits have also changed in the last century. Less time is spent preparing and eating food. Instead we are opting for more ready-made fast-foods, with ever increasing intake of fatty and sweet foods. Children often have less space to play in and are more restricted in where they are allowed to roam. Through the various media outlets a new culture of living with fear is forming. No longer are we aware of just what happens in our own small communities; we're now told about events anywhere on the planet. With increased urbanization and greater densities of population, space for play in the form of parks and our own gardens is in decline. We no longer allow our children to play unsupervised and so trips to parks and open areas become less frequent. In schools, less time is spent in physical activity for fear that children will be hurt and because of the greater emphasis on academic pursuits.

A significant increase in body fat can affect normal reproductive function in both sexes, not to mention further reducing the time spent in physical activity. In many cases, it also creates a poor self-image and may lead to avoiding social contact.

At this current time, Toronto has the distinction of being the most ethnically diverse city in the world. Over half of Toronto residents, including one of the authors and one of our children, were not born in Canada. There are more languages spoken and more ethnic backgrounds per capita than any other location on the planet. This is one of the best things about the city. It makes the concept of a global village particularly tangible to those of us who live in Toronto. However, it also means that we and other residents of the city have numerous family and social ties to the rest of the world. Because of relatively convenient and inexpensive methods of travel, we travel outside the country on a regular basis. Now like other travellers we like to bring back souvenirs of our trip. Unfortunately, such mementos of our trip might include a number of viruses that we contracted while travel. And during our trip back we may share those viruses with other travellers on the aeroplane and then the residents of our city when we return. In a cosmopolitan city such as Toronto, such actions have a negative effect, but can, paradoxically, have a positive effect in that they lead to natural immunity in part of the popoulation. Over the last few decades, we have witnessed the rapid expansion of diseases such as HIV and Lyme disease, for example, throughout a number of regions of the world. In the last few years we have seen the rate at which SARS and H1N1 viruses have spread. Human populations, in other words, are facing similar conditions to those we discussed in Chapter 8 in captive animals. We are limited by our planet and now have the potential to be subjected to diseases that arise anywhere on it.

10.4   Living with stress

Unpredictability and stress are an increasingly dominant part of our world. We have always lived with it. In the past there has always been disease, wars, famine and geological disasters. Perhaps now, with our technological inter-vention and increased awareness, we have a greater ability to combat and deal with the stressors at hand. Moreover, new generations have, in the past, proved to be particularly adaptable to the environment they have inherited. Early exposures to stress, as we described in earlier chapters, can lead to increased wariness and, therefore, avoidance of stressful events. Even if such early experience does not lead to an epigenetic alteration of the physiological stress system, then rearing under conditions that the parents found stressful may not necessarily be stressful to the children – a type of stress inoculation, so to speak. And even with the epigenetic reprogramming that could occur through high levels of maternal stress and deprivation, this mechanism need not be perma-nent. Numerous advances in cognitive therapy and other psychiatric devel-opments have shown promise for the treatment of such conditions.

We cannot predict with any great accuracy what will happen to human society in the next couple of generations. We are even less able to predict what will happen in the next two hundred years. However, we can predict, with a certain amount of confidence, that humans will probably be faced with challenges that we have not faced in the past. Will our future be the same as our past? Probably not. Each new social and technological development spurs us on into a new evolutionary pathway. New reproductive technologies will be developed that will probably enable us as a species to continue.

10.5   Summary

Humans possess a number of attributes that make our perception of stress unique to our species. These attributes include the development of our brain and body, long gestation and child-rearing periods and our society. The global human population is increasing at an unprecedented rate, and has started to challenge the availability of resources such as living space, employment, energy reserves and, increasingly, nutrition. Our society protects us from a number of stressors, but ironically creates a new set of social stressors that we must contend with. In the most technologically developed societies there has been an increase in neurodegenerative diseases, mood disorders and obesity. Many of these conditions stem from our lifestyles.

Bibliography

Chapter 1

Aloe, L., Bracci-Laudiero, L., Alleva, E. *et al.* (1994) Emotional stress induced by parachute jumping enhances blood nerve growth factor levels and distribution of nerve growth factor receptors in lymphocytes. *Proceedings of the National Academy of Sciences, USA*, **91**, 10440–10444.

Brooks, D.R. and McLennan, D.A. (2002) *The Nature of Discovery. An Evolutionary Voyage of Discovery*, University of Chicago Press, Chicago, IL.

Cooper, S.J. (2008) From Claude Bernard to Walter Cannon. Emergence of the concept of homeostasis. *Appetite*, **51**, 419–427.

Dawkins, R. (1989) *The Selfish Gene*, 30th Anniversary Edition, Oxford University Press, Oxford.

Lovejoy, D.A. (2005) Neuroendocrinology, in *An Integrative Approach*, John Wiley and Sons, Inc., New York.

Margulis, L. and Sagan, D. (2002) *Acquiring Genomes: A Theory of the Origins of Species*, Basic Books, New York.

McEwen, B.S. (1998) Stress, adaptation, and disease. Allostasis and allostatic load. *Annals of the New York Academy of Sciences*, **840**, 33–44.

Sapolsky, R.M. (1992) *Stress, the Aging Brain, and the Mechanisms of Neuron Death*, The MIT Press, Cambridge, MA.

Sterling, P. and Eyer, J. (1988) Allostasis: a new paradigm to explain arousal pathology, in *Handbook of Life Stress, Cognition and Health* (eds S. Fisher and J. Reason), John Wiley & Sons, New York, pp. 629–649.

Selye, H. (1950) *Stress*, Acta Inc., Montreal.

Selye, H. (1977) *The Stress of My Life*, McClelland and Stewart, Toronto.

Sex, Stress and Reproductive Success, First Edition. David A. Lovejoy and Dalia Barsyte.
© 2011 John Wiley & Sons, Ltd., Published 2011 by John Wiley & Sons, Ltd.

Chapter 2

Belsham, D.D. and Lovejoy, D.A. (2005) Gonadotropin-releasing hormone: Gene evolution, expression and regulation. *Vitamins and Hormones*, 71, 59–94.

Cobb, M. (2006) *The Egg and Sperm Race*, Pocket Books, London.

MacLean, P.D. (1952) Some psychiatric implications of physiological studies on fronto-temporal portion of limbic system (visceral brain). *Electroencephalography and Clinical Neurophysiology*, 4, 407–418.

Rogers, L. (2001) *Sexing the Brain*, Columbia University Press, New York.

Romeo, R.D. (2005) Neuroendocrine and behavioural development during puberty: A tale of two axes. *Vitamins and Hormones*, 71, 1–25.

Roseweir, A.K. and Millar, R.P. (2009) The role of kisspeptin in the control of gonadotropin secretion. *Human Reproduction Update*, 15, 203–212.

Rossner, S. (2007) Paul Pierre Broca. *Obesity Reviews*, 8, 277–292.

Stevenson, T.J. and MacDougall-Shackleton, S.A. (2005) Season- and age-related variation in cGnRH-I and cGnRH-II immunoreactivity in house sparrows (*Passer domesticus*). *General and Comparative Endocrinology*, 143, 33–39.

Topaloglu, A.K., Kotan, L.D., and Yuksel, B. (2010) Neurokinin B signaling in human puberty. *Journal of Neuroendocrinology*, 22, 765–770.

Chapter 3

Brown, R.E. (1994) *An Introduction to Neuroendocrinology*, Cambridge University Press, Cambridge.

Calabrese, E.J. (2008) Converging concepts: adaptive response, preconditioning and the Yerkes-Dodson Law are manifestations of hormesis. *Ageing Research Reviews*, 7, 6–20.

Lovejoy, D.A. (2006) The corticotrophin-releasing hormone family of peptides, in *Handbook of Biologically Active Peptides* (ed A. Kastin), Elsevier Science Inc., New York.

Lovejoy, D.A. (2009) Structural evolution of urotensin-I: Retaining ancestral functions before corticotropin-releasing hormone evolution. *General and Comparative Endocrinology*, 164, 15–19.

Nesse, R.M. (2009) Explaining depression: neuroscience is not enough, evolution is essential, in *Understanding Depression* (eds C.M. Pariante, R.M. Nesse, D. Nutt and L. Wolpert), Oxford University Press, Oxford, pp. 17–36.

Noyes, R. Jr and Hoehn-Saric, R. (1998) *The Anxiety Disorders*, Cambridge University Press, Cambridge.

Pepels, P. (2005) *Corticotropin-Releasing Hormone (CRH) and Neuroendocrine Regulation of the Stress Response in the Tilapia (Oreochromis Mossambicus)*, Radboud University of Nijmegen, Nijmegen.

Chapter 4

Breedlove, S.M. (1992) Sexual differentiation of the brain and behavior, in *Behavioral Endocrinology* (eds J.B. Becker, S.M. Breedlove and D. Crews), MIT Press, Cambridge, MA, pp. 39–70.

Goldberg, G. (1992) Premotor systems, attention to action and behavioural choice, in *Neurobiology of Motor Programme Selection* (eds J Kien, C.R. McCrohan and W. Winlow), Pergamon Press, Oxford, pp. 225–249.

LeVay, S. (1993) *The Sexual Brain*, MIT Press, Cambridge, MA.

Rotzinger, S., Lovejoy, D.A. and Tan, L.A. (2010) Behavioural effects of neuropeptides in rodent models of depression and anxiety. *Peptides*, **31**, 736–756.

Schulkin, J. (1999) *The Neuroendocrine Basis of Behavior*, Cambridge University Press, Cambridge.

Young, J.Z. (1978) *Programs of the Brain*, Oxford University Press, Oxford.

Chapter 5

Alonso-Alvarez, C., Bertrand, S., Devevey, G. *et al.* (2006) An experimental manipulation of life-history trajectories and resistance to oxidative stress. *Evolution*, **60**, 1913–1924.

Barsyte, D., Lovejoy, D.A. and Lithgow, G.J. (2001) Insulin signalling pathway mediated actions on heavy metal resistance and longevity in *Caenorhabditis elegans. FASEB Journal*, **15**, 627–634.

Denver, R.J. (1997) Environmental stress as a developmental cue: corticotropin-releasing hormone is a proximate mediator of adaptive phenotypic plasticity in amphibian metamorphosis. *Hormones and Behaviour*, **31**, 169–179.

Guillette, J. Jr and Crain, D.A. (1999) *Environmental Endocrine Disruptors*, Taylor and Francis, New York.

Harman, D. (1965) The free radical theory of aging: Effect of age on serum copper levels. *Journal of Gerontology*, **20**, 151–153.

Lampert, K.P. (2008) Facultative parthenogenesis in vertebrates: Reproductive error or by chance? *Sexual Development*, **2**, 290–301.

Lane, N. (2002) *Oxygen: The Molecule that Changed the World*, Oxford University Press, Oxford.

McCormick, S.D. (2009) Evolution of the hormonal control of animal performance: Insights from the seaward migration of salmon. *Integrative and Comparative Biology*, **49**, 408–421.

Munakata, A., Amano, M., Ikuta, K. *et al.* (2007) Effects of growth hormone and cortisol on the downstream migratory behaviour in masu salmon, *Oncorhynchus masou. General and Comparative Endocrinology*, **150**, 12–17.

Thomas, J.A. (1995) Gonadal-specific metal toxicology, in *Metal Toxicology* (eds R.A. Goyer, C.D. Klaasen and M.P. Waalkes), Academic Press, San Diego.

Watts, P.C., Buley, K.R., Sanderson, S. *et al.* (2006) Parthenogenesis in Komodo dragons. *Nature*, **444**, 1021–1022.

Westring, C.G., Ando, H., Kitahashi, T. *et al.* (2008) Seasonal changes in CRF-I and urotensin I transcript levels in masu salmon: correlation with cortisol secretion during spawning. *General and Comparative Endocrinology*, **155**, 126–140.

Chapter 6

Alcock, J. (1979) *Animal Behaviour: An Evolutionary Approach*, 2nd edn, Sinauer Associates, Sunderland, MA.

Bonduriansky, R. (2009) Reappraising sexual coevolution and the sex roles. *PLoS Biology*, **7**, e1000255.

Darwin, C. (1856) *Origin of Species*, 6th edn, Thomas Y. Crowell and Co., London.

Zihlman, A.L. (1997) Natural history of apes: Life-history features in males and females, in *The Evolving Female* (eds M.E. Morbeck, A. Galloway and A. Zihlman), Princeton University Press, Princeton, NJ.

Chapter 7

Allegrucci, C., Thurston, A., Lucas, E. and Young, L. (2005) Epigenetics and the germline. *Reproduction*, **129**, 137–149.

Anway, M.D. and Skinner, M.K. (2006) Epigenetic transgenerational actions of endocrine disruptors. *Endocrinology*, **147**, S43–S49.

Breuner, C. (2008) Maternal stress, glucocorticoids and the maternal/fetal match hypothesis. *Hormones and Behavior*, **54**, 485–487.

Champagne, F.A. (2008) Epigenetic mechanisms and the transgenerational effects of maternal care. *Frontiers in Neuroendocrinology*, **29**, 386–397.

Horsthemke, B. and Ludwig, M. (2005) Assisted reproduction: the epigenetic perspective. *Human Reproduction Update*, **11**, 473–482.

McGowan, P.O., Sasaki, A., D'Alessio, A.C. *et al.* (2009) Epigenetic regulation of the glucocorticoid receptor in human brain associates with childhood abuse. *Nature Neuroscience*, **12**, 342–348.

Mueller, B.R. and Bale, T.L. (2008) Sex-specific programming of offspring emotionality following stress early in pregnancy. *Journal of Neuroscience*, **28**, 9055–9065.

Chapter 8

Broom, DM. and Johnson, KG. (1993) *Stress and Animal Welfare*, Chapman and Hall, London.

Dantzer, R. and Mormede, P. (1983) Stress in farm animals: A need for reevalution. *Journal of Animal Science*, **57**, 6–18.

Krkosek, M., Gottesfeld, A., Proctor, B. *et al.* (2007) Effects of host migration, diversity and aquaculture of sea lice threats to Pacific salmon populations. *Proceedings of the Royal Society B*, **274**, 3141–4149.

Cubitt, K.F., Winberg, S., Huntingford, F.A. *et al.* (2008) Social hierarchies, growth and brain serotonin metabolism in Atlantic salmon (*Salmo salar*) kept under commercial rearing conditions. *Physiology and Behaviour*, **94**, 529–535.

Muir, J. (2005) Managing to harvest? Perspectives on the potential of aquaculture. *Philosophical Transactions of the Royal Society B*, **360**, 191–218.

Stallings, C.D. (2009) Fishery-independent data reveal negative effect of human population density on Carribean predatory fish communities. *PLoS One*, **4**, e5333.

Wilmut, I. and Highfield, R. (2006) After Dolly, in *The Promise and Perils of Human Cloning*, W.W. Norton and Company, London.

Chapter 9

Billig, H., Chun, S.Y., Eisenhauer, K. and Hsueh, A.J.W. (1996) Gonadal cell apoptosis: hormone-regulated cell demise. *Human Reproduction Update*, **2**, 103–117.

Hikim, A.P.S. and Swerdloff, R.S. (1999) Hormonal and genetic control of germ cell apoptosis in the testis. *Reviews of Reproduction*, **4**, 38–47.

Krysko, D.V., Diez-Fraile, A., Criel, G. *et al.* (2008) Life and death of female gametes during oogenesis and folliculogenesis. *Apoptosis*, **13**, 1065–1087.

Markström, E., Svensson, E.C., Shao, R. *et al.* (2002) Survival factors regulating ovarian apoptosis – dependence on follicle differentiation. *Reproduction*, **123**, 23–30.

McCord, J.M. and Edeas, M.A. (2005) SOD, oxidative stress and human pathologies: a brief history and a future vision. *Biomedicine and Pharmacotherapy*, **59**, 139–142.

Paul, C., Melton, D.W. and Saunders, P.T.K. (2008) Do heat stress and deficits in DNA repair pathways have a negative impact on male fertility? *MHR-Basic Science of Reproductive Medicine*, **14**, 1–8.

Print, C.G. and Loveland, K.L. (2000) Germ cell suicide: new insight into apoptosis durng spermatogenesis. *Bioessays*, **24**, 423–430.

Sikka, S.C., Rajasekaran, M. and Hellstrom, W.J.G. (1995) Role of oxidative stress and antioxidants in male infertility. *Journal of Andrology*, **16**, 464–468.

Tilly, J.L. (1998) Molecular and genetic basis of normal and toxicant-induced apoptosis in female germ cells. *Toxicology Letters*, **102**, 497–501.

Webb, R., Garnsworthy, P.C., Gong, J.-G. and Armstrong, D.G. (2004) Control of follicular growth: Local interactions and nutritional influences. *Journal of Animal Science*, **82**, E63–E74.

Chapter 10

Kevles, B. (1986) *Female of the Species*, Harvard University Press, Cambridge, MA.

Kikusui, T., Winslow, J.T. and Mori, Y. (2006) Social buffering: relief from stress and anxiety. *Philosophical Transactions of the Royal Society B*, **361**, 2215–2228.

Hutchinson, J.S.M. (1993) *Controlling Reproduction*, Chapman and Hall, London.

Lovejoy, C.O. (1993) Modeling human origins: are we sexy because we are intelligent or intelligent because we are sexy? in *The Origin and Evolution of Humans and Humanness* (ed T. Rasmussen), Jones and Bartlett, Boston, MA, pp. 1–28.

Maclean, P.D. (1990) *The Triune Brain in Evolution*, Plenum Press, New York.

Ridley, M. (1993) *The Red Queen*, Penguin Books, London.

Zihlman, A.L. (1997) Women's bodies, women's lives: an evolutionary perspective, in *The Evolving Female* (eds M.E. Morbeck, A. Galloway and A. Zihlman), Princeton University Press, Princeton, NJ.

Glossary

Adenohypophysis	The anterior lobe of the pituitary gland.
Adrenocorticotrophic hormone	The principal peptide hormone from the anterior pituitary gland that regulates the release of the glucocorticoids from the adrenal cortex or inter-renal gland.
Adrenalectomy	Surgical removal of the adrenal gland.
Agonist	A receptor ligand that activates the receptor.
Alarm reaction	The first stage of Selye's General Adaptation Syndrome. The alarm reaction includes the perception of the stressor and the initiation of the stress response.
Alarming stimuli	The stimuli or stressors that induce an alarm reaction.
Allostasis	The adaptation of the organism to greater challenges, such as those brought on by stressors.
Amygdala	A region of the limbic brain associated with sensory input and motor output integration, and the processing of emotions and memory.
Androgens	A group of steroid hormones primarily associated with male-associated differentiation and reproductive activity, although the hormones are also found in females. Includes testosterone, androstenedione, dihydroxytestosterone and others.
Angiotensin	Peptide hormones associated primarily with diuresis. Includes angiotensin I and angiotensin II which are processed from angiotensinogen.
Anoxia	Situations where no oxygen is available.
Antagonist	A receptor ligand that inhibits the action of the receptor.

Sex, Stress and Reproductive Success, First Edition. David A. Lovejoy and Dalia Barsyte.
© 2011 John Wiley & Sons, Ltd., Published 2011 by John Wiley & Sons, Ltd.

Antrum	The fluid-filled cavity that surrounds the ovum in a mature follicle.
Arcuate nucleus	A region of the ventral hypothalamus implicated in the regulation of growth and feeding.
Asexual reproduction	Reproduction where there is no fusion of the pronuclei. Includes fission, budding and various types of parthenogenesis.
Autonomic nervous system	Component of the nervous system that includes the sympathetic and parasympathetic nervous systems.
Basal lamina	A layer of extracellular matrix components, secreted by epithelial cells, which separates epithelial cells from deeper layers.
Basement membrane	See basal lamina.
Basomedial hypothalamus	A region of the hypothalamus at the bottom of the brain orientated toward the middle region.
Bed nucleus of the stria terminalis	A major output pathway of the amygdala.
Blood-testis barrier	A barrier composed of cells and extracellular matrix proteins that separates the seminiferous tubules from the bloodstream.
Budding	A process of asexual division where a clone of the cell or organism begins as a smaller outpocketing of the mature organism.
Carbohydrates	Simple and complex sugar and sugar polymers such as glycogen and starch.
Chemosynthesis	Synthesis of bioactive compounds that employs chemicals instead of photosynthesis.
Chordates	Animals with notochords at some point in their development.
Chromaffin cells	Cells of the adrenal or inter-renal glands that synthesize glucocorticoids.
Cingulate gyrus	Part of the limbic system situated in the medial aspect of the cortex, above the corpus collosum.
Circadian rhythms	Biological rhythms that last about a day.
Coelom	A fluid-filled cavity formed within the mesoderm.
Cyprinid fishes	Fishes that belong to the order Cypriniformes, which includes goldfish, carp and zebrafishes.
Cytokine	Peptide factors released by cells associated with the inflammatory response.

Depression	A clinical condition characterized by profound sadness, lack of motivation, disturbances in sleeping and eating patterns.
Epididymus	A region of the male aminote reproductive system that connects the efferent ducts of the testes to the vas deferens.
Fallopian tubes	Ciliated tubes that lead from the ovary to the uterus in female mammals.
Fission	A type of asexual cell division in which the cell splits into two identical daughter cells.
Fitness	Animal fitness can be defined as success at reproduction and the transfer of genetic material into the progeny.
Free fatty acid	A carboxylic acid with a long unbranched aliphatic chain ranging from 4 to 28 carbons utilized for a number of metabolic and signalling processes in organisms.
General Adaptation Syndrome	A term coined by Hans Selye to describe the organism's response to stress. It involves three main stages: alarm stage, stage of resistance and stage of exhaustion.
Genotype	The complete complement of genes within an organism.
Genotypic sex determination	Genes present that define the sex, for example XY or ZW.
Gluconeogenesis	Formation of glucose from non-carbohydrate molecules.
Glyocogen	Storage form of glucose in animals.
Glycogenolysis	Breakdown of glycogen into glucose.
Glycogen synthase	The enzyme responsible for the synthesis of glycogen from glucose.
Granulosa cells	Cells that surround the ovum that act as nurse cells to provide required nutrients and growth factors to the developing ovum.
Hagfish	The phylogenetically oldest group of chordates that possess a notochord but not a vertebrae.
Hippocampal formation	A group of structures in the brain associated with the hippocampus.
Hippocampus	A region of the forebrain associated with memory formation and learning.
Hydra	A tiny organism related to jellyfish, sea anemomes and corals and belonging to the phylum Cnideria.

Hypoxia	Low oxygen conditions.
Intersex	Individuals that have elements of both males and females but in some situations may not be functional as either.
Invertebrate	Animals without backbones.
Lampreys	The phylogenetically oldest group of animals with backbones.
Leydig cells	Testosterone-producing cells of the testes which are found outside of the seminiferous tubules.
Ligand	A molecule that binds to a receptor or other binding protein.
Limbic system	A group of structures of the brain that include the hippocampus, amygdala, parts of the thalamus and limbic cortex that are associated with memory, emotions and learning.
Lipolysis	Metabolic breakdown of fats.
Lordosis	A stereotypic reproductive and sexual behaviour elicited by rodents. Lordosis involves the lowering and arching on the back in preparation for mounting. It is typically shown by females but may be shown by males in the presence of a dominant male.
Median eminence	A vascularized region at the base of the brain that receives input from neurosecretory cells of the hypothalamus to regulate pituitary function.
Menopause	The time when menstrual periods cease.
Metazoa	Multicellular animals.
Mullerian duct	Paired ducts found in the embryo. In females they form the fallopian tubes, uterus and cervix. The ducts are lost in males.
Nephrotome	A primitive form of the kidney.
Neurosecretory cell	A neuron modified for the secretion of hormones into the vascular system.
Neurotrophic factor	Hormones and other chemical signals that have a positive effect on growth on neural cells.
Niche	The functional role a species has within an ecosystem.
Organum vasculosum of the lateral terminalis	A circumventricular region of the forebrain that does not possess a blood–brain barrier.
Orthologue	The same gene or protein found in different species.

Oviducts	The passageway from the ovaries to the outside of the body in non-mammalian vertebrates.
Panic disorder	An anxiety disorder characterized by recurring and severe panic attacks.
Parahippocampal gyrus	Cortical region that surrounds the hippocampus.
Paralogue	Genes or proteins that have arisen as a function of a gene duplication event.
Parasympathetic nervous system	Part of the autonomic nervous system associated with growth and maintenance functions.
Paraventricular nucleus	A nucleus of the hypothalamus that secretes corticotrophin-releasing factor, oxytocin and vasopressin.
Parthenogenesis	An asexual reproductive process that does not involve fusion of the pronuclei.
Parturition	Birth of the progeny.
Pathogens	Disease-bearing organisms.
Phenotype	The morphology of an individual.
Pheromones	Hormones that are secreted outside of an organism and perceived by individuals of the same species.
Phobia	An irrational and persistent fear of certain environmental events. These may include situations, people, animals, inanimate objects or activities.
Post-traumatic stress disorder	An anxiety disorder that develops after exposure to severe trauma.
Primordial sex cells	Immature sex cells.
Pronuclei	Nuclei of haploid cells (i.e. gametes).
Propeptide	Unprocessed part of the peptide hormone after the cleavage of the signal peptide.
Prostate gland	Exocrine gland of the male reproductive system in most mammals. Produces an alkaline fluid that constitutes about 25% of seminal fluid.
Protozoa	Single cell organisms.
Seminal vesicles	Exocrine gland of the male reproductive system that contributes about 60% of seminal fluid.
Seminiferous tubules	Region of the testes where sperm cells are produced.
Septal nuclei	A cluster of cell groups located in the septum, a region found between the two lateral ventricles of the telencephalon.
Sex reversal	Sex change from male to female, or female to male.

Sexual reproduction	Reproduction that involves fusion of the pronuclei of male and female.
Stage of exhaustion	The third and final stage of Selye's General Adaptation Syndrome. It includes the sum of all non-specific physiological reactions that occur as a result of a prolonged exposure to stressful stimuli where adaptations can no longer be maintained.
Stage of resistance	The second stage in Selye's General Adaptation Syndrome. It is associated with the increased resistance to a particular stressor and may in some cases be associated with a decreased resistance to other stressors.
Striatum	A subcortical region of the forebrain and major input region of the basal ganglia.
Suprachiasmatic nucleus	A nucleus of the hypothalamus associated with circadian rhythms.
Sympathetic nervous system	Part of the autonomic nervous system associated with arousal.
Tectum	Top of the midbrain, above the cerebral aqueduct.
Tegmentum	Bottom of the midbrain below the cerebral aqueduct.
Thecal cells	Cells surrounding the outside of the granulosa cells of the ovary that produce progesterone and the precursors for oestradiol.
Vertebrates	Animals with backbones.
Wolffian duct	An embryonic duct that connects the primitive kidney to the cloaca. In both males and females this duct will develop into parts of the urinary bladder and in males it develops into the epididymus, vas deferens and seminal vesicle.
Xenobiotics	Biological agents found outside of an organism that affects the physiology of that organism.
Zygote	A fertilized sex cell or the union between two sex cells.

Index

Sex, Stress and Reproductive Success, First Edition. David A. Lovejoy and Dalia Barsyte.
© 2011 John Wiley & Sons, Ltd., Published 2011 by John Wiley & Sons, Ltd.